Fracking and the Rhetoric of Place

Environmental Communication and Nature: Conflict and Ecoculture in the Anthropocene

Series Editor: C. Vail Fletcher, University of Portland

This interdisciplinary book series seeks original proposals that examine environmental communication scholarship. In the Anthropocene era, the period during which human activity has become the dominant influence on climate and the environment, the need for highlighting and re-centering nature in our worldviews and policies is urgent, as collapsing ecosystems across the globe struggle to survive. Topics might include climate change, land use conflict, water rights, natural disasters, non-human animals, the culture of nature, ecotourism, wildlife management, human/nature relationships, food studies, sustainability, eco-pedagogy, mediated nature, eco-terrorism, environmental education, ecofeminism, international development, and environmental conflict. Ultimately, scholarship that addresses the general overarching question "how do individuals and societies make sense of and act against/within/out of nature?" is welcomed. This series is open to contributions from authors in environmental communication, environmental studies, media studies, rhetoric, political science, critical geography, critical/cultural studies, and other related fields. We also seek diverse and creative epistemological and methodological framings that might include ethnography, content analysis, narrative and/or rhetorical analysis, participant observation, and community-based participatory research, among others. Successful proposals will be accessible to a multidisciplinary audience.

Recent Titles in This Series

Fracking and the Rhetoric of Place: How We Argue from Where We Stand, By Justin Mando

Water, Rhetoric, and Social Justice: A Critical Confluence, Edited by Casey R. Schmitt, Theresa R. Castor, and Christopher S. Thomas

Environmental Activism, Social Media, and Protest in China: Becoming Activists over Wild Public Networks, By Elizabeth Brunner

Natural Disasters and Risk Communication: Implications of the Cascadia Subduction Zone Megaquake, Edited by C. Vail Fletcher and Jennette Lovejoy

Critical Environmental Communication: How Does Critique Respond to the Urgency of Environmental Crisis?, By Murdoch Stephens

Natural Disasters and Risk Communication: Implications of the Cascadia Subduction Zone Megaquake, Edited by C. Vail Fletcher and Jennette Lovejoy

Communicating in the Anthropocene: Intimate Relations, Edited by C. Vail Fletcher and Alexa Dare

Communicating the Climate Crisis: New Directions for Facing What Lies Ahead, By Julia B. Corbett

Fracking and the Rhetoric of Place

How We Argue from Where We Stand

Justin Mando

LEXINGTON BOOKS
Lanham • Boulder • New York • London

Published by Lexington Books
An imprint of The Rowman & Littlefield Publishing Group, Inc.
4501 Forbes Boulevard, Suite 200, Lanham, Maryland 20706
www.rowman.com

86-90 Paul Street, London EC2A 4NE

British Library Cataloguing in Publication Information Available

Library of Congress Cataloging-in-Publication Data

Names: Mando, Justin, 1982- author.
Title: Fracking and the rhetoric of place: how we argue from where we stand / Justin Mando.
Description: Lanham : Lexington Books, [2021] | Series: Environmental communication and nature : conflict and ecoculture in the anthropocene | Includes bibliographical references and index. | Summary: "Fracking and the Rhetoric of Place investigates the rhetorical strategies of speakers on hydraulic fracturing in order to understand how places shape and are shaped by citizens as they engage in their democracy. Analysis offers scholars of place-based rhetoric and environmental communication a heuristic approach to studying their own sites"—Provided by publisher.
Identifiers: LCCN 2021034341 (print) | LCCN 2021034342 (ebook) | ISBN 9781793620873 (cloth) | ISBN 9781793620880 (epub)
Subjects: LCSH: Hydraulic fracturing—Pennsylvania—Public opinion—Case studies. | Hydraulic fracturing—Social aspects—Pennsylvania—Case studies. | Rhetoric—Political aspects—Pennsylvania—Case studies. | Public speaking—Case studies. | Environmental policy—Pennsylvania—Citizen participation—Case studies. | Risk communication—Pennsylvania—Case studies. | Communication in the environmental sciences—Pennsylvania—Case studies. | Place (Philosophy)—Case studies.
Classification: LCC TN871.255 .M36 2021 (print) | LCC TN871.255 (ebook) | DDC 338.2/728509748—dc23
LC record available at https://lccn.loc.gov/2021034341
LC ebook record available at https://lccn.loc.gov/2021034342

Contents

Acknowledgments

I must first acknowledge all people who stand up to speak out in public for the good of all. This book is about enacting citizenship through public rhetoric; the people who devote their time to matters of the common good deserve respect, even when we do not agree on policy. At a time when people all over the world, and, especially, in the United States are divided ideologically, I hold hope that citizens will continue searching for common ground with those they oppose. Through this project, I have learned how often untrue are the stereotypes that circulate on both side of those who engage in environmental issues. May we see through these flat depictions of people and learn about each other through our shared representations of place and the convictions that motivate them.

I offer the deepest gratitude to David Kaufer for his guidance through this project from my earliest days at Carnegie Mellon attempting to understand rhetoric of place. Dave and I shared the experience of being "lost in the woods." Thanks to his advice at critical junctures and his confidence in me, I was able to lead us out of a state of confusion and into the clarity I hope to have brought with this book. I also want to acknowledge the meaningful guidance and critical feedback that Linda Flower and Barbara Johnstone provided. Every time I entered her office, Linda found ways to push my thinking on whatever topic I presented well beyond where I was at when I had entered. She fostered creativity in my work and helped me to always consider the stakes. She is just the kind of professor that I strive to be, a person who sees the potential in her students and cares deeply enough to make sure she brings that potential out. Barbara has had the most impact on my project by modeling for me the habits of a true expert analyst. I learned more collaborating with her on our co-authored paper than I did on any other project during my time at Carnegie Mellon. Her attention to detail and systematic approach to analysis will always be the gold standard I strive for in my own work.

I also owe special thanks to Professors James Wynn, John Oddo and Ludmila Hyman. All three made me feel as though their doors were always open to chat about the difficult balance of work and family life. They are the type of people, not just professors, who I emulate. My friends Derek Handley and Ana Cooke have offered much needed support through the earlier days of this project and continue to be inspirational. My admiration is limitless. To my colleagues at Millersville University, especially Nicole Pfannenstiel, Kerrie Farkas, Caleb Corkery, William Archibald, Jill Craven, Kim McCollum-Clark, and Dena McEwan, I thank you for your support through the many phases of this project. You have given me a wonderful academic home. I thank Dean Ieva Zake for her encouragement and Millersville's Faculty Grants Committee for providing me a course release to complete the final stages of this book.

Thank you to Jimmy Guignard, author of *Pedaling the Sacrifice Zone*, for your help in the later stages of this book manuscript. Your book has been an inspiration and your confidence in my work has been invaluable. Let us now go explore more of the PA Wilds.

I also thank James Guerriero for his friendship and mentorship in learning the places where the stakes of hydraulic fracturing are highest. Equal gratitude goes to all the Crowfooters, Rob Doran, John Doran, Pat Thomas, Tom Jagger, and Jarret Kassan. Michael Garrigan deserves my thanks for showing me around the Susquehanna Riverlands and other places of "post-industrial wilderness." To my partners on the Board of Directors at the Lower Susquehanna Riverkeeper Association and Riverkeeper Ted Evgeniadis, I appreciate so much the opportunity you have given me to raise my own voice to advocate for the betterment of our environment.

I thank all of my students through many sections of Environmental Advocacy Writing, Science Writing, English Composition, and other courses. You have been an inspiration through your intriguing questions, your genuine concern for issues that impact us all, and your openness to the rhetorical approach.

To my mom, Donelle Davey, I cannot thank enough for being my number one supporter and always making me feel I had this in me even when the stresses of life have been at a maximum. To my dad, Jim Mando, my sister, Hillary Hirsch, and brother, Jeremy Mando, I thank you for always showing you had confidence that I could reach the end of this project. To Donald Davey, Gladys "Pug" Davey, Joseph Mando, Charlene Mando, and Scott Davey, I wish you could know how much inspiration I have found in my memories of you.

To Marika, my wife, and to my daughters, Nella and Sonia, anything I could say here to offer thanks would never scratch the surface of how important you have been to me and how much I thank you for inspiring me, for motivating me, and for giving me your love. I dedicate this book to you.

Introduction

Here or There? Hydraulic Fracturing and the Public Rhetoric of Place

Under the granite, Romanesque arches of the Allegheny County Courthouse, citizens mill about before entering the chambers. Some wear blue work shirts and weathered boots, holding across themselves a large banner made from sewn-together bed sheets that proclaims, "MARCELLUS YES!" Others, clad in green, carry banners from the Sierra Club and MarcellusProtest.org with a centered image of a gas well that reads: "FRACKING DRILLING SPILLING KILLING." Everywhere are well-defined groups of people and bobbing, hand-drawn signs hovering above the crowd: "SAVE OUR WATER"; "PA [HEART] NATURAL GAS"; "NATURAL GAS KEEPS YOUR LIGHTS ON"; "MORATORIUM ON MARCELLUS." Furrowed brows and stern glances are worn uniformly by the crowd. Above the din, shouts, slogans, and rallying cries punctuate the air. Even before the public hearing has begun, people are making their voices heard. As their clothes and banners suggest, their words are anchored to the land.

This particular public hearing took place on March 10, 2011 in Pittsburgh, Pennsylvania and was meant to gather comments on whether high-volume hydraulic fracturing—widely known as *fracking*—should be permitted within the county. At that time, southwestern Pennsylvania, which sits atop the Marcellus Shale formation, was one of many "shale plays" dramatically reshaping global energy markets as American gas production rocketed from 1 trillion cubic feet per year in 2000 to 13 trillion in 2014, coinciding with the 2014 crash in global oil prices (Raimi, 2017, p. 27).

This eye-popping increase in American production is the result of thousands upon thousands of wells drilled across the country. In 2011, just one year during the peak of the shale gas boom, Pennsylvania alone saw 1,937 wells drilled into the Marcellus Shale (Duggan-Haas et al. 2013, p. 39). This feat was accomplished by some of the 169,000 new employees of the fossil fuel related jobs that were added to the US economy between 2010 and 2012 (Gold, 2014, p. 26). Such statistics suggest that rampant natural gas extraction in Pennsylvania, other areas of the United States, and across the world is a foregone conclusion, but this is not so. Efforts by concerned citizens to halt or slow down gas development have gained traction. States like New York continue to uphold a moratorium, and regional bodies like the Delaware River Basin Commission, which represents the states of New York, Pennsylvania, New Jersey, and Delaware, prohibit fracking within the watershed. The vastly different approaches to the discovery of massive natural gas reserves between states like Pennsylvania and New York are political and rest upon differing thresholds for risk. To determine these thresholds, governing bodies need, in part, to listen to their citizens.

Citizens representing all angles of the fracking issue have taken public opportunities to voice their opinions. Many have personal stakes in the outcome beyond the impacts on their wallet. These stakes often have much to do with the proximity between fracking and their homes. A reporter for the *Wall Street Journal*, Russell Gold (2014), explains just how close many live to fracking sites: "As of 2013, more than 15 million Americans lived within a mile of a well that had been fracked in the past few years" (p. 19). In contrast to previously distant locales in Saudi Arabia, Nigeria, Russia, and elsewhere, the closeness of fracking brings the process of energy production into view. This shift has caused Americans to engage the question of where fracking should be done. It becomes a question of *here* or *there*.

The primary concern of this book is to listen to citizens respond to the question of whether to frack *here*. In doing so, we come to better understand the role of place-based arguments in public rhetoric on fracking. My purpose is not to convince you that we should or should not frack our lands, but instead to reveal how the relationship between people and the land on which we live instills place with value, which facilitates the role of place as a rhetorical resource. This emerges in the rhetorical efforts of citizens on both sides of the fracking debate as they locate their arguments in place. These efforts belong to both those who support and oppose fracking, as they all have a stake in place. As Meera Subramanian (2019) explains in her listening tour of "red states" facing climate change as ". . . the shared lament between the climate-stubborn Right, the Green New Deal Left, and the majority who inhabit the space in between: the fundamental shared experience of the land shifting beneath our feet" (p. 21). Indeed, the people engaged in the controversy of fracking can testify to the changes taking place on what had seemed like stable ground. In this sense, this book is not just about public engagement with fracking, but also how we speak for our changing environment from wherever we call *here*.

Studying fracking in the American context, researchers of rhetoric engage with a case where public and private interests collide in public forums.

Private business may convince private citizens to lease their land, but the impacts of these decisions are not confined to those involved in these private transactions. Due to the public impact, scientists, environmentalists, industry representatives, politicians, landowners, business owners, healthcare workers, educators, and many other citizens converge to determine whether the risk is worth the reward. This rhetoric reverberates across the land. The discussions of one community influence the available arguments of another.

While fracking is a widespread industry that demands national and international conversation, local governments and citizens play a key role in regulating it. This, in part, is why county-level hearings become such contentious sites. Whether those who hold the institutional power to decide on hydraulic fracturing can reach consensus, often through inter-governmental hearings, is a complicated matter affected largely by how questions are reframed to avoid criticism (Buttny, 2015) and whether they can look past the industry's strategic framing of fracking as patriotic, science-based and safe (Matz & Renfrew, 2014). These questions enter into all levels of society, from state-level debates between experts (Buttny, 2019) to international engagement through various media (Metze & Dodge, 2016; Matthews & Hansen, 2018). At a local level, proximity to fracking operations impacts the availability of "not-in-my-backyard" or NIMBY appeals, while opposition to such appeals are often framed as a fight against the liberal left (Dokshin, 2016) and an opportunity to reinvigorate "Rust Belt identities" (Rich, 2016). Yet the complexities go much deeper than partisan politics, as individuals work to establish their position on an issue enmeshed in place-based attachment and understandings of what belongs (Guignard, 2015; Kroepsch, 2016). One of the primary sites of this discourse are the many public hearings held throughout shale gas regions.

By analyzing public testimony, this book offers insight into the rhetorics of place used by citizens. The strategies uncovered here offer rhetoricians new perspectives on the importance of place in public argument and an analytical approach to researching the rhetorics of place in other sites. Learning that supporters and opponents of fracking all rely on place-based appeals, but that they craft these appeals quite differently, can inform our own attempts to create impactful rhetoric. This has pedagogical implications too, as we teach others to discover their own relationship to the places around them and that their words can shape the world. This book, in its broadest sense, is about listening to citizens in institutions of ordinary democracy while attempting to change the minds of those around them through rhetoric that is both ensconced in and shapes the places where they live, the places they visit, and even places that they have known only vicariously.

Therefore, the first objective of this book is to offer a heuristic approach to the analysis of the rhetorics of place that focuses on language in use through the mixed methods of discourse analysis and corpus-based rhetorical analysis. In the current historical moment, rhetoricians often focus their gaze on a broadening array of texts and have vast quantities of material ripe for analysis at our fingertips, such as computer-aided techniques that can process large corpora and offer insights that would be impractical or unnoticed by the human reader. However, computer-assisted corpus studies can lack in the details that can be gleaned from the close reading of individual texts. My approach provides analysts a way to pair the wide-angled lens of corpus analysis with the fine-toothed comb of discourse analysis. This book uses utterances about place as its unit of analysis and the inclusion criteria for the corpora upon which this book is based.

The second objective of this book is to theorize the rhetorics of place as citizens use accounts of the physical world to engage with public environmental controversy. As a rhetorician, I offer insights into the ways language shapes and is shaped by the places we inhabit. Place-based arguments are a topos of public rhetoric that draw power from the land itself, but this is not a unidirectional process like a tree that draws nutrients from the soil. Our linguistic configurations of place shape places themselves. Rhetoric is material. Rhetoric changes our physical world and the fracking debate is an exemplary case of this phenomenon.

The third objective of this book is to support the role of language in use as we continue to theorize rhetoric and new materialism. Rhetoricians have successfully extended our reach to the non-linguistic world and become captivated by the rhetorical force of the nonhuman and the nonverbal. These are undoubtedly important ventures for our field. However, we may view this as a swing of the pendulum that at its opposite extreme, during the origins of rhetorical criticism, closely reads the single speeches of powerful white men. The more equitable, inclusive, and diverse rhetorical agents of a new materialist perspective is a move in the right direction, but in the absence of language our analyses lose some amount of rigor and replicability. In the space of the highly theoretical, interpretation is at a premium. This book calls for rhetoricians to pair investigations of the material with linguistic texts that can be submitted to rhetorical analysis. Language gives us the concrete evidence we need to support our interpretations and to trace the steps taken to get from phenomena to interpretation.

The fourth objective of this book is to turn theory into practice as we teach students, our citizens-in-training, how to engage with local spaces through place-based appeals. The scope of our lives is increasingly broadened beyond the local. Our experiences are now so frequently virtual that connections to physical place may escape more people than ever before. At the same time, as these words are written during the COVID-19 pandemic, people have turned to the natural world in the form of parks, trails, rivers, lakes, mountains, and other sites of outdoor experience to cope with the limitations of a global lockdown. The time we spend in the natural world is valuable for our mental health and, as this book shows, to our lives as citizens. When we pay close mind to the places we share with our fellow citizens, understanding that these are sites of literal common ground, we gain access to powerful tools that can be used to shape the world through our public rhetoric. This skill does not come without practice, both in paying close attention to places and in articulating the meaning of those places to others. The closing section of this book takes the theoretical insights about the rhetorics of place and rearticulates them into practical exercises that can help us see the places we inhabit more deeply and clearly. Once we have seen clearly, we can strategically communicate from and about places to shape the future of the land.

VALUES THAT GROW FROM THE GROUND

All social controversies involve competing value systems. Land-use controversies, such as hydraulic fracturing, are no different. In a controversy like fracking, values grow from the ground. Stakeholders articulate their differing values through specific connections to the land itself. Killingsworth and Palmer (1992) explain: "In all ethical problems, we must consider the rules for community formation, but in environmental disputes, we must additionally understand how the disputants construct their views of the natural or nonhuman worlds" (p. 4). Because it is through language that much of our understanding of the physical world is constructed, rhetoricians are well suited for this type of work. Rhetoricians, discourse analysts and other English and communication scholars have been increasingly interested in how the land is constructed, how values are placed into these linguistic constructions of human and nonhuman worlds, and how these representations can clash.

This book adds to this conversation insights from a case that covers the planet but that manifests itself in ordinary people's backyards. When the tanker trucks start rolling into town and drill rigs are on the horizon, some people take up arms and others are only too glad to lend a shovel. Their reasons for resisting an influx of fracking or cheering it on are complex and go well beyond the rhetoric from which these rationales spring. Whether these people have long histories in their places or young families, they have vested interest in the prosperity of their piece of the earth and value leaving behind a cleaner world than they received. How they articulate their values is directly connected to the physical place they inhabit. While a rhetorical study may not disinter all of an individual's reasons for resisting or supporting a process like fracking or recreate the full context, through rhetorical analysis of a community's public rhetoric we can better understand how we landscape our experiences in place to fit our rhetorical purposes. Most interestingly, common rhetorical strategies emerge in public hearing speeches independent of formal strategizing. People speak to the best of their ability and topoi of place-based rhetoric emerge.

Discourse scholars contribute to environmental debates through their attention to the functions of language in particular contexts as well as the connections between symbols and identity. Steve Depoe (1998), past president of the International Environmental Communication Association, appealed for the importance of studying environmental communication well before his organization's founding in 2011 by arguing that rhetoricians can make significant contributions in this field. Though the categories he highlights are present in other discourses, Depoe explains that environmental communication importantly serves both constitutive and instrumental functions. Nature, he claims, is "symbolically constructed and enacted through discourse" (p. 436). Depoe is supported by Cantrill and Oravec (1996), who assert that talk shapes the physical world, especially when regarding the environment. Rhetoricians are poised to intervene in such issues, potentially shaping their outcomes.

The rhetorician's toolkit proves useful to providing analysis and even a potential solution to what at first seems an "irreconcilable conflict." An example of the power of rhetorical intervention comes from Mark P. Moore's (1993) work on the spotted owl's use as a synecdoche in the Pacific Northwest. Moore shows that to environmentalists, spotted owls are an indicator species, but to the timber industry, they are a scapegoat that causes job loss. Through the owl synecdoche, Moore suggests a different framing of the problem that re-presents ecology and economy as compatible. In this configuration, the owl prompts faster action because inaction will only make the problem harder to solve (p. 271). May we find rhetorical interventions that reshape the gas controversy?

Beyond the symbolic and its constitutive effect, Depoe (1998) also emphasizes the instrumental value of the work of environmentally minded rhetoricians and other environmental communicators who "intend to manage and care for the physical spaces around them" (p. 436). Critical works by rhetoricians and discourse analysts enact some of this "care" for the places around us, such as an analysis of the *Sydney Morning Herald's* representation of an island nation threatened by climate change that challenges marginalizing discourse aimed at Tuvaluans (Farbotko, 2005) or the globalizing and neoliberal influence of Alberta (Canada) place-branding materials as they relate to bituminous sands (Takach, 2013). The instrumental value of environmental communication research can also serve to inform communities about rhetorical strategies for advocacy in issues where place is contested. Place transcends its status as setting and becomes an integral component of argument.

Central to a rhetor's identification efforts are experiences within places, especially nature, that cannot be communicated. The Emersonian perspective that we "know more from nature than we can at will communicate" (Emerson, 1986, p. 19) resonates throughout environmental rhetoric and communication scholarship, as in Oravec's (1981) study of John Muir's use of the sublime, the work of Donal Carbaugh (1996, 1999, 2007) and Carbaugh and Cerulli (2013) on the connection between place and environmental knowing. On this Emersonian way of knowing nature, Carbaugh (1996) explains,

> This, I believe, is one reason we walk on trails, climb peaks, or, for other reasons, visit sites of land use controversies, or landfills, oil spills, and so on. As a result of being a part of nature's spaces, we "know" more and thus might work to say something else, something better, about our places. (p. 40)

In spite of our inability to communicate fully what we have witnessed, the contact with nature and sites of land-use controversies that Carbaugh describes drive an attempt to use language to shape the future perception of these natural places.

Whether a coal controversy in Yorkshire, England (Usher, 2013), a conservation effort in an Iowan watershed (Gottschalk-Drushke, 2013), the community-based discourse of Hispanic youth in the American Southwest (Milstein et al., 2011), or the speeches that describe interactions with the fracking regions of Pennsylvania analyzed in this book, people draw on their own particular sense-of-place and their experiences in places as they use language. Through this, they define their specific "ecocultural premises," that is, what is "integral to a community's place-based relationship that brought them into being" (Milstein et al., 2011, p. 488). Stated differently, interactions with places and contests over what those places mean shape the values and identities of the people that live within them. These understandings of place within communities are always in flux, continually constructed and reconstructed. This is why public hearings are such an important site to study the rhetoric of place: citizens must construct the value of place in discourse alongside their fellow community members, building in aggregate robust representations of their world.

Carbaugh and Cerulli (2013) explain that environmental discourse is always "doubly placed" in that it is "always located in place, and, it shapes our sense of places" (p. 7). Place affects and is affected by discourse. Studying the rhetoric of people in situ as they describe their world, its locations, relationships, important features, and inhabitants can help us examine this reflexive relationship between place and discourse. "We can do so by listening for the discursive devices people use to make place, nature and dwelling there explicit, such as place-names, depictions of places including stories, ethno-physical nomenclature, and spatial deixis" (p. 17). In other words, we must view the language of place as strategic, rhetorical, and revealing of the values we both find and place in the land. Carbaugh and Cerulli's call to focus on place in environmental issues has been answered by scholars across disciplines who take this social constructionist stance.

Analyses of environmental communication from a social constructionist perspective can be a powerful tool to ground policies, positions, and perspectives from the bottom-up. Geographers Bickerstaff and Walker (2013) take a rhetorical and discourse analytic approach to air pollution, a notoriously difficult-to-communicate risk due to its (most often) invisibility. These geographers survey work in their field to better understand how the public constructs pollution, polluters, polluted places, and polluted people. They find that each of these categories is knowable, as it is "embodied and centrally embedded in place" (p. 59). For instance, in one case (Bush, Moffatt, & Dunn, 2001), proximity to pollution sources determined whether or not people felt that pollution was a problem. Despite having no evidence for a lower content of pollutants in their own air, they believed that poorer communities close to the source of pollution were at risk while their own community was not.

While Bickerstaff and Walker (2013) reveal how proximity affects perceptions of air pollution from those distant and close to the issue, Cantrill and Masluk (1996) take a different perspective on proximity and place in their study of how people of differing economic status perceive industrial activity in the American Rocky Mountains (p. 80). In their comparison of the perceptions of upper-class owners of vacation homes and more lower-class permanent residents, they found that those with higher socio-economic status viewed industrial progress more positively. Because residents with less financial security are unable to move away from development, they become more negative: they "cannot escape the threat in their backyard" (p. 84). These studies have influenced the central role that *proximity* plays in the investigation of the dynamics between the here and there in this book; analyzing what it means to dwell somewhere, including the difference between residing on land as a visitor or permanent resident, facilitates our understanding of the rhetoric of place.

Proximity has also been shown to affect the news value of media reporting (van Dijk, 1988; Bednarek & Caple, 2014). In addition, it also affects how events are reported (Johnstone & Mando, 2015). The work of Piotr Cap's (2013) and others on *proximization theory*, which builds on Paul Chilton's (2005) *discourse space theory*, advances our understanding of how proximity is constructed through language that creates a sense of impending threat. Blommaert et al. (2003) look at how physical proximity to a refugee center affects people's discourse in their examination of people who live varying distances from the center. They find that close proximity is related to stronger degrees of certainty, distrust of outside media, and a hunger for information about the topic of controversy. Proximization also increases the threat of invasive species, such as Asian carp, which is being thwarted from entering the Great Lakes (Mando & Stack, 2019). These studies show how the rhetorical construction of proximity, especially the proximity of threat, can prompt drastic action.

Proximity relies on the related concept of *deixis*, as deictic elements connect the rhetor, their audiences, and the place. Fahnestock (2011) explains that these elements "include all those referring words that identify the people and relevant objects immediately present to the speaker and hearer" (p. 325). Especially important are the seemingly insignificant adverbials of place like "here" and "there." In the "Gettysburg Address," Lincoln repeats the word "here" six times in three sentences to rhetorically connect his audience to those who died on the battlefield, as well as their shared duties to their nation. In this sense, deixis creates involvement by asking audiences to envision the scenes they depict (Prasch, 2015, p. 253). Deixis can be used to connect rhetors and audiences, but when pointing to what is *here* metaphorical barriers can be erected between what is *there*. Often, we assign negative values to what is distant (Cap, 2018, p. 383). Much of the controversy that is the focal point of this book is entangled in the interplay between the proximate and the distant, including the reasoning behind how these deictic elements are construed.

Scholars who explore proximity-based reasoning, even those outside of environmental issues, illuminate the relationship between proximity, perception, and rhetoric. By attending to how people construct place and considering these constructions as constitutive of identities and values, we can gain an understanding of the risks people face and the values that matter to them, allowing us to better learn how to use place as an interface that informs communication efforts both within and outside. It is for this reason that I focus on citizens at public hearings. Their testimonies and use of the rhetorics of place influence the construction of their identities, whether they are farmers, hikers, scientists, or drillers. Juxtaposed to these people who are rooted in the land, we also consider the role of the scholar who engages with these discourses from afar. We may learn from the people who have their boots on the ground, but what about the scholar's boots? Are they to remain clean? Where do we dig in and how? What must we consider as we engage citizen's public discourse?

Scholars in rhetoric and environmental communication now call for us to investigate rhetoric more holistically, in a systems-focused way that does not privilege the symbolic over the material. Edbauer (2005) introduces us to this approach through her concept of *rhetorical ecologies*. To view rhetoric ecologically, we must regard it as an interweaving of the symbolic and the material that requires a kind of engagement through which researchers seek out patterns, tensions, and overlaps across a range of broadly defined texts.

For instance, reading the rhetoric of a national park could involve analyzing trail maps, the trails themselves and the land they bisect, the humans and nonhumans who visit the trails, and the researcher's own embodied experience (Senda-Cook, 2013). Such ecological views of rhetoric ask us to make way for the material alongside the material.

This view of rhetoric as ecology follows Dana Cloud's (1994) *third way*, which emphasizes the role of the ideology critic in working with the "*persuasive force of rhetoric in history*" (italics in original, p. 158). Cloud pushes against what she sees as two flawed notions of materiality in discourse: the first, championed by McGee (1990), McGuire (1990), and Ono and Sloop (1992), states that "rhetoric, language, and culture transcend and determine material relations of power (idealism)." This egocentric position privileges human symbol-making and is, according to Cloud, idealistic, assuming rhetoric, language, and culture to be outside of the material. The second, more relativistic position puts us in the same bind. Cloud summarizes it through the works of Laclau and Mouffe (1985), McKerrow (1989, 1991), Baudrillard (1975), and Foucault (1980): ". . . that rhetoric, language, and culture constitute reality and that therefore there are no ontological or epistemological grounds for moral or political critique" (p. 158). Notwithstanding Cloud's critique that such a stance removes the rhetorician's ability to meaningfully address moral and political issues, the problem is that both views privilege the human and leave no room for the nonhuman (Rogers, 1998). Rogers refers to them as constitutive theories of discourse where reality is socially constructed through human symbol-making. He states that these theories "hold that nature has no *meaning*, no reality, beyond that assigned through symbols" (p. 244). In some sense, Cloud addresses Rogers's concern, as her view of ideology does not see rhetoric's role as deterministic or all-encompassing. She encourages us to leave space for the nonhuman and material, which leads to a more ecological view of the world.

Some studies have attributed rhetorical agency to nonverbal and/or nonhuman action, though I argue that they tend to be limited. Two such efforts focus on the material impact of walking where one does not "belong." Both Robert Topinka's (2012) study of walkers in commercial spaces designed for automobile traffic and Natasha Seegert's (2014) article that follows Bruno the Bear's atypical journey out of the Alps and into Germany help us see how rhetorical concepts elucidate material practices. Topinka argues that rhetorical tropes deployed through material rather than symbolic means allow for a novel approach to rhetorical invention (p. 82). In other words, a person who walks across a space designed for vehicular traffic or a bear who walks beyond its typical range performs an embodied act of resistance. While the person in Seegert's analysis ignored the hegemony of urban planners, Bruno the Bear "challenged boundaries of what is permissible" and "disrupted our human sense of control of the landscape" (p. 88). Both studies show the appeal of applying rhetorical concepts to material practices. Their work expands the range of what we can do as rhetoricians and opens fruitful lines of inquiry.

However, such studies do raise questions about the purview of rhetoric and how far we may productively stretch our discipline. My particular concern is that they overlook foundational beliefs for rhetoricians regarding audience and intentionality. Are the drivers who are inconvenienced by a walker in a shopping center the audience for Topinka's subversion? Or is it the gardener whose flowers are trampled? Is the teenager who walks through the shopping center because they are too young to drive performing the same act of resistance? Can a bear challenge "boundaries or what is permissible, and normal" (p. 88)? Without considering audience or intention, are we on solid ground when we apply rhetorical concepts to the material? As these studies show, such applications of rhetorical theory can be fascinating and informative, but I question whether others might replicate their results. In their creative interpretation of material practice, they claim a high degree of significance for what they study. However, opposite interpretations could likely be supported. From my perspective, we should continue to study the material with more replicable rhetorical and discourse analyses of symbolic representations through language use.

Such a middle ground between the material and symbolic has been advanced by Salvador and Clarke (2011) in their article on the *weyekin principle*; it has also been put into practice by others in our field (Ackerman et al., 2016; Rai & Gottschalk-Druschke, 2018; Guignard, 2015, Senda-Cook, 2013). The weyekin principle grew from ethnographic research on a Nez Perce wolf introduction program (Salvador & Clarke, 2011) that led Salvador and Clarke to "*an embodied critical rhetoric the researcher attends to the corporeal experience of the nonhuman world so as to articulate the symbolic-material tensions obscured by predominant systems of meaning*" (italics in original, p. 248). At first glance, such an approach appears to be ethnography that casts a wider net, but Salvador and Clarke emphasize the importance of becoming in tune with the nonhuman entities present, likening it to a spiritual connection. The process starts by "listening beyond symbols," which "requires more than a senseless walk in nature. It demands close attention, rigorous observation, and embodied presence" (251). While this foray into the material, nonsymbolic, and nonhuman may seem to raise similar worries about rigor and replicability, Salvador and Clarke present the "weyekin principle" as complimentary to—not a replacement for—rhetorical analysis (pp. 251–252).

Scholars who blend such embodied engagement with more traditional forms of rhetorical analysis offer more insight into the ecology of place.

Guignard's (2015) *Pedaling the Sacrifice Zone* renders a sophisticated textual analysis that is deepened by his immersion in and physical attunement with place. After the death of a friend to a cycling accident in the forest of a fracked region of Pennsylvania County, Guignard reflects, "What this tells me is that good memories and bad come together to create my sense of this place, and that lived history gets inscribed onto the landscape the same way that words get written on a page" (p. 127). How can we as rhetorical analysts and environmental communication scholars learn this lived history without, simply put, living it? Must we put down the texts to encounter a rhetorical ecology's "biological attributes, rhythms, and circulation" (Ackerman et al., 2016, p. 77)? I do certainly support and encourage such embodied approaches to rhetorical analysis when combined with textual analysis, but we must realize that the *lived history* of a place cannot be hastily apprehended.

In this study of place-based rhetoric derived from citizen testimony, I often see the speakers as experts in their place to a degree that I could never attain without having lived alongside them. What I find on both sides of the fracking issue are representations of the "weyekin principle" in the words of people who have listened to the land a long time. Although in many cases, they have not listened comprehensively to the land, animals, plants, and water, as Salvador and Clarke would stipulate, they too demonstrate a degree of ecological listening. By listening to the people who have inhabited the land, we learn how they represent their voices and their land, remembering that they, not us, have known it intimately for decades. With this in mind, I write this book to represent the voices of those who have listened and now speak.

DATA-DRIVEN APPROACHES TO RHETORICAL ANALYSIS

Inspired by empirical and data-driven approaches to rhetorical analysis, I use discourse analysis as a rhetorician's methodology (Johnstone & Eisenhart, 2008; Mando, 2016; Cloud, 2019; Thompson, 2013; Oddo, 2014). In an article in *Written Communication*, John Oddo (2013) argues for the suitability of discourse analysis as a tool for rhetoricians and provides readers with a "micro-rhetorical toolkit." Oddo dispels worries that attention to method can constrain creativity (McKerrow, 1989, p. 102) and that discourse analysts are "fixated on the formal features of texts" and ignore context (p. 238). He sustains that discourse analysis does not stifle creativity by explaining the heuristic, not formulaic, approach that discourse analysis (DA) encourages. In terms of its formal approach, Oddo cites Hopper, responding that discourse analysts are "less interested in linguistic forms and more interested in the 'micro-rhetorical' *functions* of language" (p. 238).

The shift toward DA in rhetoric studies may have been in part instigated by Ernest Bormann's (1972) work on everyday talk. Bormann set a precedent when he developed his theory of fantasy and rhetorical vision in an analysis of small group meetings "as a rhetorical critic might analyze the text of a public speech" (Bormann, 1972, p. 396). Bormann was one of the first to analyze everyday talk from a rhetorical perspective, which today is more common. As Oddo cites in his article in *Written Communication*, some rhetoricians already use DA (Fahnestock, 2011; Cheng, 2012; Cramer, 2011; Dunmire, 2011; Barton & Stygall, 2002; Huckin 1992, 1997, 2002; Huckin, Andrus & Clary-Lemon, 2012). Scholars who are particularly interested in the rhetoric of place/space have been drawn to work on deixis, which was first introduced by linguist Karl Buhler (1939) in *Theory of Language: The Representational Function of Language* and recently made accessible for rhetoricians by Jeanne Fahnestock (2011) in *Rhetorical Style*. Fahnestock's distinction between imaginary and immediate deixis has proved useful in public address scholarship on place in presidential speeches (Prasch, 2015). My work is another step toward fruitful combinations of discourse and rhetorical analysis.

Editors of collections that bring together rhetoric and discourse analysis have made progress toward demonstrating the compatibility of these disciplines. Johnstone and Eisenhart's (2008) *Rhetoric in Detail: Discourse Analyses of Rhetorical Talk and Text* and Tracy, McDaniel, and Gronbeck's (2007) *The Prettier Doll: Rhetoric, Discourse and Ordinary Democracy* are notable works of this type. Johnstone and Eisenhart's collection includes work by rhetoricians who use discourse analysis as a theory and/or method, while Tracy et al.'s collection brings discourse analysts and rhetoricians together to provide complimentary perspectives on one case study. Tracy et al. explain that in both disciplines "scholars . . . share a conviction that how individuals and societies go about constructing representations matters and requires focused analysis. Both believe that looking at texts carefully, taking account of what exactly is said or written, is consequential" (p. 13). The authors then explain that both rhetoricians and discourse analysts are interested in texts that are strategic in nature, "in which creators, whether subtly or obviously, artfully or crudely, attempt to generate knowledge and opinion, motivate action and attitude, and overall induce others to cooperate" (p. 13). This highlights the similar focus of both disciplines on text creation as a strategic endeavor made understandable through qualitative textual analysis that also considers context.

While both disciplines share similarities, Tracy et al. (2007) highlight what many believe bifurcates rhetoric and discourse analysis:

> By and large rhetoricians assume a critical, evaluative stance in relation to their object of study, whereas discourse analysts give greater weight to description.

> From the rhetorical point of view, the evaluation in discourse traditions may seem so "lite" as to make it hard to taste. For discourse analysts a carefully worked up description is essential before moving to critique. From the discourse analytic point of view, rhetoricians are prone to rush to judgment, often inattentive to important particulars of the scene. (p. 14)

From a general perspective, the difference holds between critical rhetoricians and descriptive discourse analysts. However, with the rise of Critical Discourse Analysis led by Norman Fairclough (1989), Ruth Wodak (1989), and Teun van Dijk (1990) among others, discourse analysts have had new tools for critique and evaluation. Movement toward DA as a method for rhetorical analysis, demonstrated by Johnstone and Eisenhart, has provided sturdier textual grounding for rhetoricians. In some regards, rhetoricians and discourse analysts are moving closer together. This book advances rhetoricians' use of DA as a method by operating within a site of ordinary democracy, following work by Tracy, McDaniel and Gronbeck. I use a combination of discourse and rhetorical analysis to provide insights about public hearing discourse on the topic of hydraulic fracturing.

PUBLIC RHETORIC IN PUBLIC HEARINGS

As this project combines rhetorical analysis and DA, I have selected a site to investigate that incorporates the strengths of both rhetoricians and discourse analysts. There are shortcomings of this site that both rhetoricians and discourse analysts would note, which I will also discuss in this section. I study transcripts of public hearings that deal with hydraulic fracturing and natural gas extraction at a local level, through city or county governmental hearings that invite public comment. From a discourse analyst's perspective, my focus on the talk of ordinary citizens and corpus-based approach are familiar territory. For rhetoricians, my focus on democratic participation and public address are strong suits. Public hearings provide an opportunity to draw from the strengths of rhetoric and discourse analysis.

Transcription is typically important to discourse analysts as the process of entextualizing language entails choices that impact the way language is interpreted. As explained by Rodney H. Jones (2011), "To be engaged in creating texts about reality is to be engaged in creating reality" (p. 11). When an analyst selects the discourse features to include in a transcript, they make certain analysis possible. For instance, transcription that includes extralinguistic features like pauses and repairs can provide important information for conversation analysts about power relations and identity formation. Court reporter transcripts, however, not transcriptions of the hearings, are sufficient for this study because my aim is to understand the rhetorical use of place. My unit of analysis is the place reference and its relationship to larger argument structures. Although much of the original hearings (some of which I have viewed in video) is lost in the court reporters' transcripts, as Jones says "too often analysts have taken a 'deficit' attitude towards entextualization, lamenting how much of the 'original' interaction was 'lost in the transcription'. . . ." Jones goes on to explain that in transcripts we are not attempting to represent "truth," but rather "analytic utility." A transcript's merits should be judged on its "ability to help us answer the questions we have about human communication and social interaction, not the degree to which [it] 'resemble[s] reality'" (p. 20). In this regard, the analytic utility of court reporters' transcripts is sufficient for my purposes.

For rhetoricians, my focus on transcripts to provide insight on how citizens use place as an argumentative resource may be considered divorced from context. It is true that this book does not recreate the context of each hearing I analyze. I have studied the media reports that precede or follow these hearings, but they are not the object of my analysis. In addition, I have not analyzed the "rhetorical ecology" (Edbauer, 2005) of each text by assembling a corpus of interconnected social texts representing a variety of genres. In response to such criticism, I say that the purpose of my study is to understand particular discourse moves within the context of the hearing itself. I do believe it would be illuminating to study the circulation of discourse and how place-based appeals factor into that circulation, but I see this book as groundwork for later explorations on broader context.[1] Though I admit that contextual factors would likely offer new insights, I do heavily weigh the physical location of the primary hearing in my work and its location within Allegheny County, Pennsylvania, the Marcellus Shale region, and the United States. As place is a dominant trope in this study, relative location plays a large role in my work. However, I tend to this by considering how proximity or distance is created through language.

I have most extensively analyzed the 110-page court reporter's transcript of the March 10, 2011 Allegheny County public hearing on Marcellus Shale. This hearing serves as the primary text for chapters 2 and 3. In chapter 4, I expand to include seven hearings from 2008–2014 located in Pennsylvania, New York and Illinois (see table in chapter 4). In all of this, I narrow my data by selecting place references. That is, any time a speaker mentions a concrete place using a place name or a relational descriptor, I consider this part of my initial dataset. For instance, if a hearing participant refers to Deer Park Lake (place name) or "a state forest ten miles northwest of here" (relational descriptor), I select this reference and note the argument it supports.

In each chapter, I select my data to interrogate a different aspect of place-based appeals. In chapter 2, I segment my data by type of place (i.e., natural, industrial, home). In chapter 3, I again use place-based appeals but focus on specific uses of place that create imaginable scenes. Through this, I develop vicarious proximity appeals as a rhetorical topos. In chapter 4, I use the vicarious proximity appeal as a unit of analysis applied to seven hearings. In these hearings, I have only selected vicarious proximity appeals, seeking an even number of 30 appeals from pro-frackers and 30 from anti-frackers.

RHETORICAL ANALYSIS AND DISCOURSE ANALYSIS

My approach to analyzing these texts has been to begin broadly, narrowing into specifics of place-based discourse while moving from a smaller to a larger corpus. I do this to gain an understanding of the primary ways place is used as an argumentative resource in one hearing, then to specify one particular way place is used in that hearing to best understand what it affords speakers. Finally, I take what I have learned and apply it to a wider set of texts to see how segments of the population use this specific appeal to place differently. The progression of my chapters follows a grounded practical theory approach (Craig & Tracy, 1995). This approach advocates examining three levels of a practice: (1) the technical level, (2) the problem level, and (3) the philosophical level. The technical level consists of rhetorical strategies and discourse practices. The problem level is "made up of the dilemmas that speakers confront in participating in the practice" (Tracy & Hughes, 2014, p. 309). The philosophical level is essentially the normative ideal that participants in a practice hold toward that practice. I limit my discussion of the philosophical level for my final chapter as I advance normative claims about the practice of participating in a public hearing on land use. In chapters 2, 3, and 4, I weave the technical and problem levels to interrogate how and why speakers use place in public hearings. In the final chapter, I reach the philosophical level and offer suggestions for future practice.

By shifting between the technical and problem levels, I have found a methodology that blends rhetorical and discourse analysis to provide me with the micro-level lens to understand the technical level in terms of genre, narrative and rhetorical strategies of deliberative argument. Broadly speaking, I use genre analysis to understand public hearing discourse. I look at both the centripetal and centrifugal forces of this genre, that is, what makes it a genre and the differences within it. Through computer-aided analysis and hand-coding, this work can be described as move analysis (Swales, 1990) that is primarily focused on place at the lexico-grammatical level. I also view these texts from a narrative analysis perspective as I focus on setting and scene (Labov & Waletzky, 1967; Labov, 1972). Public hearing speeches that involve place are short narratives that implicate identity formation (Bamberg & Georgakopolou, 2008).

All this is to say that I bring multiple analytic lenses to my texts and I work with different units of analysis to best understand the way place is used as an argumentative resource. This helps me understand place from micro- and macro-rhetorical levels, which, when taken together, provides insights into the rhetorical strategies of public hearing participants. Such an approach is grounded in the texts and offers rhetorical scholars interested in place an empirical method that can complement more abstract approaches. I intend this blended method to be explicit and replicable so that others may use this book as a guide to their own sites of investigation.

COMPUTER-AIDED CORPUS ANALYSIS

The entirety of this analysis, but specifically in chapter 4, is informed by computer-aided methods of rhetorical analysis. In doing so, I join a growing number of rhetoricians who employ these methods. One of these is Roderick Hart, whose DICTION program is designed to analyze texts for tone. Discussing his program, Hart explains four key assumptions: "(1) people use words to do things; (2) they use them in varying proportions; (3) audiences react to these deployments cognitively, socially, and emotionally; and (4) they are guided by implicit understandings of rhetorical form when doing so" (Hart, 2015, p. 153). Hart's program, along with other computerized tools, makes the "varying proportions" of words in a text readily apparent. This can be gleaned through hand-coding, but what computerized tools make possible is the analysis of large corpora. Computers make it possible for us to understand the moves that are common to larger populations (Hart 155). Computer-aided analysis also provides these proportions for myriad categories of language. Hart explains, "Because computer programs like DICTION attack a text from forty or fifty different angles simultaneously, each reading becomes a separate reading but also a mutually implicative reading, telling the user when a text conforms to, or deviates from, a set of norms" (p. 155). This set of norms often comes from reference dictionaries that are built from corpora representative of the English language.

I use David Kaufer's and Suguru Ishizaki's DocuScope as my primary tool for computer-aided rhetorical analysis. In several articles, Kaufer and his co-authors describe the function and usefulness of a digital "environment" that aids in rhetorical analysis. In "Textual Genre Analysis and Identification," Kaufer et al. (2005) start with a detailed look at close reading popularized by I.A. Richards. This entails two modes of looking—looking through and looking at—a text. The textual analyst must perform both of these functions, examining content as well as content distribution. In order to classify the genre of a text, both of these processes are necessary. Kaufer et al. set out to create DocuScope to identify and isolate the elements in a text that activate particular genres in readers.

Rather than rely on single-word dictionaries such as DICTION, as previous programs have done, DocuScope categorizes strings of language and assigns a frequency variable determined by the relative frequency of the string in relation to a large corpus of texts. The reason for this is that Kaufer et al. wanted to make transparent the "analytical procedures" employed by writers and therefore recover their rhetorical strategies. In this, the program serves not just as a genre-classifying tool, but as an aid to rhetorical analysis.

To understand how the program functions, Kaufer et al. offer a useful analogy. They describe the writer as engaging in "rhetorical scrabble"; instead of 26 letters to shape into words, they have 15,000—100,000 words to form into strings that are given meaning when used in a particular context. Their program identifies strings, not words, so that ambiguities between words can be distinguished. For instance, Kaufer et al. classify uses of the word "smear" that have positive or negative associations in different contexts. After a painstaking coding process, a large textual input process where they load various corpora into the program, and a quality control and testing phase, the program has millions of recognizable strings and can aid in the classification of texts into genres. Kaufer has tested the abilities of his program by teaming up with Robert Hariman (2008), who recently classified political discourse into four categories based on his own close reading, and the program successfully identified these categories and helped substantiate Hariman's claims.

RHETORICAL ANALYSIS AND PUBLIC ARGUMENT

This book separates hearing speakers into "pro-frackers" and "anti-frackers" to study the dyadic relationship to this issue vis-à-vis the primary argumentative positions available to public hearing participants. Despite my dialogical focus of pro and anti, this controversy can certainly be read as a *polylogue* of the "many-to-many" subjects and audiences that engage with this issue (Aakhus and Lewinski 2017, p. 179). However, the public hearing format invites clear boundaries between those who support and those who oppose. The rhetorical situation calls for rhetors to state clearly their position, but there is plenty of room for nuance, even in a three-minute speech. By taking a dyadic approach, this book seeks this nuance and reports on surprising ways that rhetors navigate the complexity of what appears to be a classic rhetorical situation featuring a single context, consistent set of constraints, exigence, a speaker, and their audience (Bitzer, 1968). We may view these speeches in this way, but we must also consider that these public speakers inhabit multiple rhetorical situations at once (Kjeldsen 2008, p. 116). Explained by Aakhus and Lewinski (2017), "As a speech travels to new audiences who take it up and react to it and demand new responses, the speech takes on a life of its own" (p. 187). In other words, these speeches do not begin and end in the public hearing. They continue to influence the many positions available to members of this public outside of the space of the public hearing. How such rhetoric circulates through publics and the process whereby publics form, grow, and share information is the subject of a long history of scholarship that informs the approach to public hearing discourse in this book.

CHAPTER SUMMARIES

Chapter 1—This chapter details the case study of hydraulic fracturing and offers the theoretical support for this work through theories of the public sphere.

Chapter 2—In this chapter, I analyze all references to places in the March 10, 2011, public hearing, focusing on both concrete and general references to place to reveal generalizable strategies of place representation through synecdoche by anti- and pro-frackers.

Chapter 3—This chapter begins my exploration of vicarious proximity appeals, derived from analysis of the March 10, 2011, public hearing. I define vicarious proximity appeals through features critical to the construction of these appeals and I use them as a unit of analysis to provide insight into the different uses of the appeal by anti- and pro-frackers. Anti-frackers tend to represent places up-close from an on-the-ground perspective while pro-frackers zoom out and provide more of an aerial view of landscape. I argue that this is done to respectively increase the sense of threat or create a sense of safety.

Chapter 4—I use computer-aided corpus analysis methods to analyze a larger dataset of vicarious proximity appeals from eight public hearings. This analysis reveals a micro-rhetorical strategy advanced via vicarious proximity appeals: the use of *apologia*. Speakers representing pro- and anti-frackers work to repair stereotypical images of their sides of the issue through place representation strategies.

Chapter 5—This chapter summarizes the findings of the book, provides commentary on the value of place-based rhetoric to public deliberation, risk communication, and rhetorical theorists. I detail how place-based rhetoric functions as an interface to global and local energy issues. Pedagogical implications of my findings are explored through the *Tiny Ecology Project*, a place-based experiential learning project.

NOTE

1. In the final chapter, I discuss the connection between my work and Jimmy Guignard's book on the rhetoric of fracking in Pennsylvania. Guignard has taken an ecological approach while I have remained focused on specific language in use. Despite this, we reach strikingly similar conclusions that support our approaches and deepen the insights that can be gained from both texts.

Chapter 1

Shaping Place with Public Rhetoric

Citizens who wish to engage in their democracies have many venues available for expressing concerns, encouraging the direction of policy, and articulating their values for the politicians who will ultimately enact policies. Alongside the formal institutions are the myriad civic acts that constitute one as a member of a public. Signing an online petition, posting about an issue on a social media site like Facebook or Twitter, and even discussing issues of common public concern among friends are acts of participatory citizenship. These acts and the discourse through which they take shape can circulate as coalitions form, arguments develop, and examples come to light that bolster individual experience into collective action.

Contact points that span from the individual to the broadest abstract notions of democracy facilitate the study of democracy in action, allowing us to navigate the myriad ways citizenship becomes enacted. These contact points, through which individuals establish common ground, embody democracy. What is the best course of action for all of us? How do we act for the common good? To drill into the way these questions are answered, I focus on *place* as a central topos of democratic rhetoric because it provides firm ground on which to observe democracy as a practice of the people. The place-based issue of hydraulic fracturing and its public hearings adjust the knobs of this investigation to a fine focal point.

When I entered this research space, it was on my own two feet as a hiker covering the trails of western Pennsylvania. My wife, Marika, and I had recently returned to my home state and were living in Pittsburgh, from where we would travel to explore the expansive trail systems and rivers that held trout and smallmouth bass. It did not take long for us to encounter the industrialization that hydraulic fracturing operations had brought to Pennsylvania. Access roads that cut into woodlands, hiking trails that had been intersected where we least expected to find heavy tanker trucks, and the traffic of industry were common sights that gave us pause. Drawn to the opposition, I soon found grassroots efforts to oppose fracking and was intrigued by the rhetorical energy being spent to raise awareness and build coalitions. At the same time, we were confronted daily by the idyllic representations of Pennsylvania in natural gas industry advertising. I found that they made a strong case for the benefits of fracking, including bringing back the blue-collar jobs that had been lost and providing a needed boon especially to the rural areas of the state. It was clear that the common ground to this issue was all around us. We hiked it, farmers tilled it, and drillers drilled it. Yet, I asked myself, how could these strongly opposed parties be heard and have their collective voices represented in the decisions that this and many other states and countries were making as natural gas and unconventional hydraulic fracturing boomed?

At this point, I started paying close attention to the contest over place, not just Pennsylvania as a site of fracking, but the individual farmer's acres, the trout angler's favorite stream, the drill worker's job site, and the birdwatcher's woodlot. As citizens like these publicly engaged with the issue of hydraulic fracturing, they continually articulated their relationship to

Figure 1.1 Tower for horizontally drilling into the Marcellus Shale in Lycoming County, Pennsylvania, USA. *Source*: Photograph by Ruhrfisch (Wikimedia Commons).

place. To me, it seemed that interpreting place-based rhetoric was the key to understanding the complexities of this particular issue and, more broadly, democracy in action. My first notion was that those who opposed fracking—unfairly stereotyped as tree huggers, hippies, the Prius-driving liberals who lived in cities and only came out to the forests to play—were the ones who couched their rhetoric in a "love of place." Yet, the Range Resources advertisements that featured farmers representing generations who had all lived and worked the land called this notion into question. Even scientists, I found, relied on place-based rhetoric. Terry Engelder, a Penn State University geoscientist and vocal hydraulic fracturing supporter, explained in *Nature*, "I grew up with the sights, sounds and smells of the Bradford oil fields in New York State. My parents' small farm was over a small oil pool, with fumes from unplugged wells in the air and small oil seeps coating still waters. . . . Like me, most people living in such areas are not opposed to drilling, it seems" (Howarth et al., 2011, p. 274). What I began to find is that place-based appeals are a ubiquitous resource in public rhetoric, but they are used differently by those who support and oppose hydraulic fracturing. As people on both sides use place-based appeals to advance deliberative argument, they contest the meaning of places in their communities and beyond. This came to particular light in community-based public hearings on hydraulic fracturing, where contests over place featured prominently in the speeches of citizens. Place-based appeals appeared to be doing a lot more rhetorical legwork than simply establishing that a speaker was from *here* and not *there.*

In addition to being a topos, a common source of rhetoric, I take the perspective that place is also a *rhetorical interface* where contextual, social and personal constraints meet in a "constructive process" (Flower, 1994, p. 44). In a public hearing, citizens use it not only to construct their arguments, but also their identities, affiliations, and values. In the process, a burgeoning local understanding of the common good and of what it means to engage as a citizen in democracy come forth. From this vantage, long-contested theories of the public sphere can be juxtaposed to actual practice. Through the shared rhetorical strategies of citizens, that is, we gain insights into how democracy works so that we can refine our theories of how it *should* work. In addition, by conceptualizing place as an interface, I have used place-based arguments as a method for dividing texts into meaningful units that reveal connections between speakers, audiences, and the rhetorical situations they enter. These lessons can be put into practice by writers who wish to advocate for places and the people connected to them. To grasp these lessons and to understand how the insights of citizens engaging in their democracies can be put to use, I should explain the issue that has brought them together in the first place.

WHAT IS HYDRAULIC FRACTURING AND WHY SHOULD WE CARE?

Pennsylvania has been particularly prominent in the national and international conversation on fracking due to the aggressive development by energy companies within the state, a familiar position due to its strong history of extraction. In spite of this focus, hydraulic fracturing is geographically widespread. Energy producers around the globe find this newly accessible and valuable resource irresistible. In their book, *Hydraulic Fracturing*, Smith and Montgomery (2015) claim fracking to be "one of the most important and enduring technologies ever developed by the oil industry"; they explain that this process has added "9 billion barrels of oil and over 700 trillion standard cubic feet (scf) of gas to U.S. reserves that would have otherwise not been economical to develop" (p. 1). For many, especially from the US standpoint, this domestic resource is welcomed as a "gift" (Duggan-Haas et al., 2013, p. 2). In his 2013 State of the Union Address, Barack Obama even touted the scale and personal impact of hydraulic fracturing when he explained, "We produce more natural gas than ever before—and nearly everyone's energy bill is lower because of it." From these perspectives, natural gas extraction has its clear benefits, but as many citizens testify, becoming energy independent has its costs. These values are being debated wherever hydraulic fracturing occurs.

Where shale gas reserves are found in North America, Europe, and, increasingly, the rest of the world, hydraulic fracturing is an issue of contention. This process can be traced back to the 1890s, when nitroglycerin was used to stimulate wells (Smith & Montgomery, 2015, p. 2); however, the process that is at the center of global controversy focuses on *unconventional* natural gas extraction. The distinction between hydraulic fracturing and unconventional natural gas extraction is an important one that separates a process with a long history from one that is relatively new. Buttny and Feldpausch-Parker (2016) explain the distinction between conventional and unconventional natural gas extraction in the "Communicating Hydrofracking" special issue of *Environmental Communication*:

> "Unconventional" in the sense that the oil or gas is not found in an underground reservoir, but must be extracted from a shale formation or coal bed using a fracturing process that injects large amounts of water, chemical propping agents, and sand at high pressure to release oil or natural gas from the formation. (Duggan-Haas et al., 2013) While fracturing of shale rock has been around for more than 60 years, it is its coupling with horizontal drilling that has made it so lucrative. (p. 289)

The crucial difference is the technology that allows drillers to reach deeper depths and, once they have reached the desired depth (around 9,000 feet), to extract in a horizontal path. This is achieved over a long and gradual curve as new drill sections are attached. Drills shift two to four degrees until reaching a horizontal orientation (Sernovitz, 2016, p. 26). With this technology, energy companies can extract more gas with fewer wells, increasing profits substantially. Because unconventional gas extraction is so lucrative, many see natural gas as a way to boost rural economies, transition away from coal, and achieve national energy independence (Buttny & Feldpausch-Parker, 2016, p. 289).

The dream of energy independence from the riches of natural gas, however, is currently in peril. Due to the extensive extraction of natural gas, prices have fallen to meet the abundant supply. The price per 1,000 cubic feet of natural gas was $8 in 2002, but fell to an average of $4 from 2009 to 2011, and continued to decline to $2.66 in 2012 (Duggan-Haas et al., 2013, p. 39). During the beginning of the COVID-19 pandemic, natural gas appeared to have a much more uncertain future. At around $1.80 per thousand cubic feet, this marked another year of the price decline that began nearly a decade ago. Yet, the factor that likely will drastically impact natural gas production, potentially irreparably hobbling the entire industry, is the precipitous drop in oil prices. In January of 2020, before the global pandemic halted travel and other oil-dependent industries, the price per barrel was around $60. In May of 2020, six weeks into quarantine, the price per barrel was just below $13.

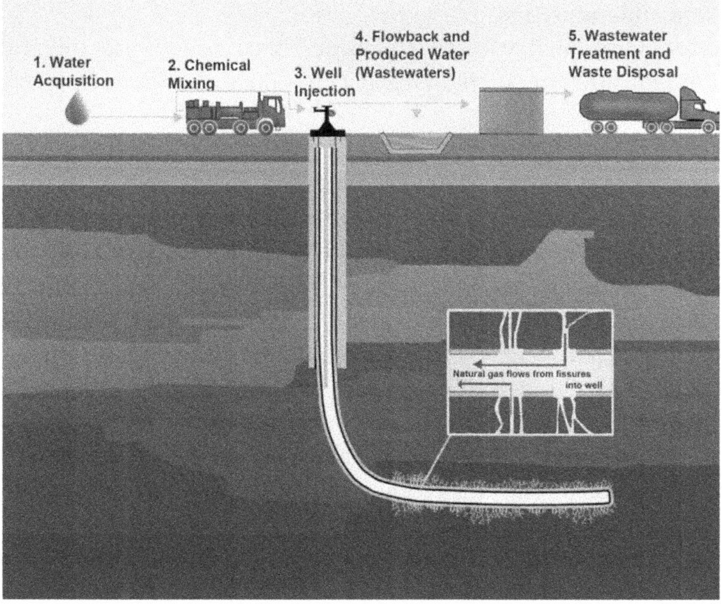

Figure 1.2 This Diagram Shows the Horizontal Orientation of an Unconventional Natural Gas Well and the Steps to Complete Such a Well. *Source*: US Environmental Protection Agency.

Prices of both natural gas and oil have since recovered in May of 2021, but this volatility may cause gas extraction to become a much less lucrative enterprise, possibly leading to the end of a long and impactful boom.

The public hearings represented in this book come from the period when gas prices were in decline, but leasing land to make profits from gas extraction remained enticing to many. The governmental intervention that took away the choice to lease from landowners who, in part, argued that this energy source was clean in comparison to oil or coal was disconcerting to many (Chen & Gunster, 2016, p. 315). When fracking occurs, others have emphasized concerns over water contamination, air pollution, community disruptions due to increased industrial activity, and seismic activity (Brasier et al., 2011). Wastewater from this process contains radioactivity from radium (Duggan-Haas et al., 2013, p. 25) and posts serious challenges regarding disposal (p. 86). These concerns are most certainly real, despite widespread claims for the safety of unconventional natural gas drilling. A review of the 1,778 peer-reviewed studies from 2009 to 2015 conducted by Concerned Health Professionals of New York & Physicians for Social Responsibility (PSR) finds that

- 69 percent of original research studies on water quality found potential for, or actual evidence of, fracking-associated water contamination,
- 87 percent of original research studies on air quality found significant air pollutant emissions, and
- 84 percent of original research studies on human health risks found signs of harm or indication of potential harm. (Physicians for Social Responsibility, 2019, p. 12)

It is important to note that "potential for harm" is different from findings of actual harm, but, as detailed by Physicians for Social Responsibility; plenty of evidence of fracking's actual negative impact exists. More evidence mounts as researchers continue to study the impacts of hydraulic fracturing, such as a 235-page Pennsylvania grand jury report that charges the Pennsylvania Department of Environmental Protection and the Pennsylvania Department of Health as failing to adequately protect the public (Supreme Court of Pennsylvania, 2020). A more specific study found that methane leaks at volumes more than twice what was previously assumed by the US Environmental Protection Agency during oil and gas production in the Permian Basin and at levels much higher than average in eleven other major US basins (Zhang et al. 2020). The leakage in the Permian Basin alone represents an energy loss enough "to supply 7 million households in the state of Texas" (p. 4). Weakening US federal regulations on gas capture and emissions brought about in 2018 contribute to this issue by making management of methane an unnecessary expense (p. 5). Medical researchers, in a study of 35,508 patients, have found asthma exacerbated by proximity to unconventional natural gas development (Rasmussen et al., 2016). These findings and those compiled by the PSR damage the common argument that a switch to natural gas from coal has made energy production in the United States a cleaner enterprise.

The issue of hydraulic fracturing is not localized to Pennsylvania, the Marcellus Shale region, or even the United States. Although the United States pioneered this technology, it is now being spread across the world. China, Russia, and Argentina are rapidly developing shale gas reserves, as are Algeria and Poland (Watts, 2011; Raimi, 2017, p. 176). While some countries develop natural gas reserves, many have imposed moratoriums and outright bans on unconventional drilling technology. France, Ireland, Spain, and Bulgaria have banned hydraulic fracturing (PSR, 2019, pp. 13–15), Germany has a moratorium until 2021, the Australian state Tasmania has a moratorium until 2025 and Victoria implemented a permanent ban in 2016 (PSR, 2019, p. 14). South Africa also resists developing their massive shale reserves (Raimi, 2017, p. 176). The European Union's Committee on Industry, Research, and Energy suggests member states resist authorizing new unconventional gas development as it "entails risks and adverse consequences for the climate, the environment and public health, and jeopardises the achievement of the EU's long-term decarbonization goal." Indeed, a study of the United Kingdom's Office of Unconventional Gas and Oil finds that citizens resist monologic presentations on behalf of natural gas; they want dialogue and democratic deliberation on the issue (Williams et al., 2017, pp. 100–101). Despite the potential for resource development, citizens in the EU and other countries who resist unconventional natural gas development have won just as they have in New York and, more recently, Maryland (PSR, 2019, p. 15) and the Delaware River Basin (Delaware River Basin Commission, 2021). This resistance is despite the financial influence that politicians have faced. According to a report by Conversation Voters of PA, The Marcellus Shale Coalition and others in the hydraulic fracturing industry from 2007 to 2018 spent $69.6 million to lobby Pennsylvania's state government. Additionally, lobbyists and others related to the industry donated $11.2 million during this same period (Zilcosky, 2018).

However, there is one important difference between the United States and international contexts for fracking: how we define mineral rights. The United States is one of the only countries on Earth that grants private ownership of mineral rights. In all countries mentioned in the previous paragraph, the direct proceeds of gas extraction go to the state, whereas US citizens individually stand to gain by choosing to extract gas from their properties. While a $50,000 leasing bonus goes straight into a US citizen's pocket, elsewhere that money is made by the state (Raimi, 2017, p. 175). In a state like

Pennsylvania, the draw for private individuals to take the risks associated with fracking is especially enticing in the many rural communities that have experienced economic strain since the last natural resource-based boom.

Pennsylvania has a long history of natural resource extraction dating back to the mid-nineteenth century, when industrial development in the form of anthracite coal mining and timber gathering (largely to support the construction of mine shafts) grew unchecked (Wilber, 2012, p. 13). During the second half of the nineteenth century, several million acres of Pennsylvania's forests were clear-cut (Brasch, 2013, p. 21) and "hillsides were stripped bare" (Wilber 2012, p. 12). In addition to coal and timber, northwestern Pennsylvania is home to the first petroleum well in the United States: the Drake's Well, which was tapped in 1858. Edwin Drake's discovery led to a rapid expansion of petroleum drilling that lasted nearly a half century (p. 14). These industries have brought cycles of boom and bust, prosperity and ruin. The human costs of this industrial expansion have been high; a documented 31,000 miners died due to anthracite mining alone from 1869 to 1999 (p. 13). Environmental devastation has also been a cost of this development, from the clear-cut zones of the timber industry to the orange streams that still flow with acid mine drainage to air quality issues resulting from steel and coke plants. Some argue that hydraulic fracturing is an opportunity we must take and that we can recover from any troubles it may bring; others claim that we cannot make the same mistakes of the past.

Embedded in the long Pennsylvanian history of extractive industries are stories of financial gain and ruin, prosperity and desperation, health and sickness, and places that have been despoiled, recovered, and despoiled again. What we must consider is how this same land and its history can be viewed so differently by those who live on it. Some people see the wonders of God's creation that is resilient to our use of its bounties. They see the rivers that now run cleaner than they have in many decades as an indication that any pollution from our use of natural resources will be washed away over time. Others see fragile natural systems that are finally restoring their balance after generations of pollution and carelessness. They believe we cannot afford to risk damage that is irrecoverable.

PUBLIC HEARINGS ON HYDRAULIC FRACTURING

Back at the Allegheny County Courthouse, people have filtered into the county council chambers and the public hearing has begun. Members of the public take turns arguing for their position. A common theme emerges; representatives of industry, members of environmental advocacy groups, scientists, health professionals, as well as people who do not identify with a group spend significant time describing their personal experiences with hydraulic fracturing. They talk of their farmlands, their hikes through state forests, visits to rural counties and small towns. People create impressions for their audience of how these places have changed due to hydraulic fracturing for good or for ill.

On July 22, 2010, the Environmental Protection Agency held a hearing in Pittsburgh that more than 1,200 people attended. Don Hopey of the Pittsburgh Post-Gazette wrote of that hearing:

> Although EPA officials told those in attendance the meeting was not about drilling policy, most of the more than 100 speakers let it be known that they oppose Marcellus Shale drilling in the state, and many shared personal stories of contaminated wells, dead farm animals and damaged health. They attributed the problems to water contamination caused by the deep gas drilling operations that are increasing quickly through much of the state. (Hopey, 2011)

Industry representatives such as Lou D'Amico, president and executive director of the Pennsylvania Independent Oil and Gas Association, provided opposing viewpoints: "Fracking is neither a new nor controversial process . . . Any controversy is based on hysteria, not facts. It's had no negative impact on groundwater anywhere it's been used" (Hopey, 2011). This federal-level

Figure 1.3 Pittsburgh, Pennsylvania, Seat of Allegheny County. *Source*: Photograph by Michael Yatsko.

hearing was one outlet for citizens on both sides of the issue, but the EPA made it clear that policy discussion was not their intent.

On the local level, around the same time Allegheny County started to have hearings that were first oriented to collect public sentiment and then later to communicate specific policies. The hearing that is the subject of the bulk of this analysis was held at the Allegheny County Courthouse on March 10, 2011. It was so well attended that speakers were asked to exit the chambers after they were done speaking to allow others to enter. The chambers were full of participants who numbered roughly two-to-one in favor of those who opposed fracking (Jansa, 2011). Most speeches were read from the podium, which was positioned on the left side of the room facing the opposite wall. Few were given extemporaneously. Standing between the audience and the county council, who faced each other, many speakers had prepared comments and concluded with additional remarks that were not read. In the middle of the room was a table with a stenographer. Media representatives, who faced the council members, sat at a raised platform. This is a rather typical arrangement in council chambers when the public gives testimony. Citizens faced their council members with fellow citizens at their back, emphasizing the power differential and the face-threatening act they perform when they rise from the citizenry to speak out (Brown & Levinson, 1987).

At public hearings, the discourse of fracking appears to be heavily skewed in terms of race. In this particular hearing, the vast majority of participants were white, with one speaker identifying as from Mumbai, India. Women and men were approximately evenly represented in the audience and among podium speakers. In the six additional hearings that I cover in chapter 4, the racial and gender make-up is strikingly similar. Despite this lack of diversity, public hearing discourse still represents an important site where competing voices converge. Although the speakers at the meetings were not particularly diverse, they gave voice to circulating discourses that reached beyond their own interests. This is not to say that the lack of diversity in public hearings is not a problem that should be addressed.

Along with racial discrepancies, researchers from Clark University have found that fracking is strongly correlated with poverty in Pennsylvania, West Virginia, and Ohio (Ogneva-Himmelberer & Huang, 2015). However, the timing, location, and cost of attending public hearings limit the representation of lower-class voices. Hydraulic fracturing hearings typically occur on weeknights (school nights), beginning at 6:00 p.m. and ending around midnight. Because impoverished families tend to live in rural parts of the county, they would have further to travel and fewer resources to cover costs of travel and childcare to allow them to attend a meeting in downtown Pittsburgh. Further, at least one study of public meetings has confirmed that older, white, male homeowners are the most typical participants in local government (Einstein et al. 2019). Although some speakers do speak for those who are not present or were unable to attend, we must remember from the onset that many important voices in this public are not present. That said, the public hearing is a mode of democratic participation that puts the public into direct contact with those who make decisions that impact the lives of a community. In addition, they are a contact zone where people challenge and are challenged by opposing viewpoints (Koban, 2019, p. 3). Some speakers do speak for those who are not present or who were unable to attend, but we must remember from the onset that many important voices in this public are not present. Public hearings can be sites of power imbalance (Koban, 2019; Buttny, 2015; Buttny & Cohen, 2007). In spite of their significant limitations, public hearings thereby continue to offer insight into the rhetorical practices of citizens as they engage with democracy. Critical attention by rhetorical scholars on public hearing discourse may also "illuminate the value structures that underpin citizen comments and reveal a new path forward for making public hearings a space where citizens have more influence on environmental decision-making processes" (Scarff, 2021, p. 2). This ground-up approach that investigates what citizens do when they participate in the democracy can inform normative approaches as well as policy.

WHAT SHOULD DEMOCRACY LOOK LIKE?

Central to American political life is the idea that citizens should work together for the common good of all. How best to accomplish such selfless acts of cooperation is an issue of disagreement on both practical and theoretical levels. The axis of this disagreement is not so much a conflict of specific values, indicating that people want different things. Instead, people tend to disagree on the hierarchical arrangement of similar values. We can agree that inclusiveness and rational argument are important for fairness in politics, but which should we value more? Should we preference the advice of technical experts over the pleas of people who have lived in a place for generations? Questions like these come to bear in a practical sense in institutions of democracy such as the local public hearing. Tradition meets innovation and manifests as a practice. Our ideas of what democratic practice should look like have been shaped and continue to be shaped by theorists of the public sphere who question how to value logical argument, inclusiveness, expertise, and secondary aims of public participation like identity (Asen, 2003) and discursive will formation (Benhabib, 1992). These public sphere scholars consider what counts as a compelling argument, who should participate, and the ultimate aim of such participation. Many of these studies are normative in nature in that they suggest what democracy should look like and raise awareness of the tensions between models of the public sphere. Though this study takes as its foundation the idea that we must study democratic practice in its local instantiations to best understand what works in democracy, normative theories provide entry points into productive lines of inquiry.

I take a discourse-focused approach to public participation to investigate how people argue over a local and land-based dispute in the public sphere. By presenting the myriad tensions of "doing democracy," our attention can be focused on what ordinary people do that aligns with normative theories of the public sphere as well as those practices that are underrepresented. I expand upon the position championed by Karen Tracy (2010), which starts with the practices of democracy as it actually exists and builds up to theory. This is in juxtaposition to models that impose a normative framework on actual practice. In essence, this perspective is inquiry-driven, looking to the phenomenon of local democratic engagement to learn how democracy works instead of asking how well people live up to our ideals of democracy. This is not to disparage normative theories of the public sphere, as they provide us with guidance and the careful consideration that can lead to novel insights and updates to more fossilized practices. However, when it comes to environmental controversy on land-based disputes, citizens still have much to teach us about the already existing norms of participation, evidence, and effective argument.

In spite of class and racial limitations, public hearings on fracking and the broader political process of gathering community members to speak on public issues rest on a model of participation that imagines citizens to be on even ground. They step forward to speak into a microphone for equal allotments of time, which tends to be three minutes. During that time, they can offer their perspective to the governing bodies before them and to their fellow citizens at their backs. Those who defend the practice of public hearings accept that the "authority . . . of the better argument" should win the day (Habermas, 1989, p. 41). People have equal time, equal space, and equal opportunity to craft a persuasive argument. The trust that reason rises to the surface rests upon the influential work by Jurgen Habermas (1989) in *Structural Transformation of the Public Sphere*, which is known as the rational-critical model of the public sphere. An important pre-cursor to a study such as mine on the deliberation over a technical process like hydraulic fracturing, Habermas's rational-critical model emphasizes the role of reason over passion. He explains that historically, as people moved toward democratic models of governance, they saw "the subjection of political domination, as a domination of human beings over human beings, to reason" (p. 128). Habermas believes publics should disregard the status of participants and bracket their differences (pp. 36–37). Ideally, this would happen in a public hearing, and hearings are operated in ways that appear to foster this bracketing by providing each participant, whether an elected official or an unknown citizen, equal time and an equal chance to voice their perspective. In practice, however, this does not happen. Speakers routinely provide personal details, tell their stories, and relate their experience in the places where they live. As a brief example from a public hearing held in 2018 before the Delaware River Basin Commission, a speaker states: "My name is Greg Lotoro. I'm from Milford, Pennsylvania. The Delaware River has always been a part of my life. Growing up, I fought to stop the Tocks Island Dam and more recently, have been working to prevent the basin from being negatively impacted by gas extraction and gas infrastructures. I'm here today to ask the DRBC to. . . ." This excerpt represents the ways that many speakers begin their public rhetoric by emphasizing their connection to place. Others form this connection not through personal history but through organizational affiliation, such as an environmental or industrial organization. Regardless, they wear their identities on their sleeves and do not attempt to persuade through solely rational argument.

Another issue is that the rational-critical model does not account for the rising need of publics to deal with issues that are highly technical where complex understanding seems a prerequisite to being persuaded by the better argument. The temptation to rely solely on technical experts for issues such as hydraulic fracturing pushes against the rational-critical model and, as theorists show, moves into dangerous territory. Habermas laments the rise of mass media culture and advertising's impact on politics (p. 219); however, in his later work (2015), *The Lure of Technocracy*, he rejects the subjugation of reason to technical expertise in the context of the European Union. Habermas and others, like Gerard Hauser (1999), reject technocratic models of the public sphere in part due to a shift in the responsibility of the public from developing a will that policy-makers must reflect in policy toward "shap[ing] the will of the governed in line with the policies to be enacted" (Habermas, 1989, p. 27). In a democratic model, policy-makers turn their constituents' will into action; in a technocratic model, policy-makers persuade their constituents that the technocrats' solution is the right one. This situation, where the public has lost touch with policy, is discussed in detail in Farrell and Goodnight's (1981) case study of Three Mile Island. They find that "the discourse at Three Mile Island seemed to be governed by a conception of the public itself as an accident, an audience extrinsic to the community of technologists, indirectly addressed only during times of expert dissensus" (p. 282). The authors blame the consequences of this accident on the inability of technical experts to communicate danger. Instead, technical experts pacified their audience, rendering them unable to engage with the issue.

In the face of a public that is powerless against technical experts, the rational-critical approach seems a decent solution. However, Nancy Fraser (1990) shows us that Habermas's model has deficiencies, too. This kind of stripped-down participation where reason rules the day is complicit with hegemonic domination, as the idea of bracketing difference excludes women, people of color, non-property owners and many others (p. 116). Bracketing difference, she explains, is not possible, as the modes of discourse in public interaction favor the dominant group: "[Public spheres] are not spaces of zero-degree culture, equally hospitable to any possible form of cultural expression" (p. 126). Because of this inequity, attention is always paid to social difference even if attempts are made to formally bracket it. Instead, Fraser believes we should thematize difference (pp. 119–120). The idea of thematizing difference has led scholars to consider alternate forms of representation in public spaces beyond the logical appeal and outside of Habermas's rational-critical model. Consider again the acts of protest and personal representation made by sign-carrying citizens outside of the Allegheny County Courthouse. This, along with the example above of a typical public hearing introduction, is evidence that citizens do readily thematize difference.

Rhetorical models of the public sphere provide a window into how citizens might thematize difference rather than value while engaging in democracy. The creative, emotive, personal, and identity-forming rhetoric that is so prevalent whenever people make and defend their perspectives in public helps us see the differences that make individuals unique. Even if we continue to value logical argument, looking at discourse from a rhetorical perspective shows a more complete picture. Before Habermas's influential work, John Dewey (1927) proposed that the quality of public engagement was tied to rhetoric. He explains:

> Presentation is fundamentally important, and presentation is a question of art . . . The function of art has always been to break through the crust of conventionalized and routine consciousness. Common things, a flower, a gleam of moonlight, the song of a bird, not things rare and remote, are the means with which the deeper levels of life are touched so that they spring up as desire and thought. (pp. 182–184)

Dewey's emphasis on "art" relates to the deliberate design of symbols that communicate to audiences beyond the cold power of logic. Likewise, Robert Asen (2004) explains that "democracy asks not for people's unlimited energy and knowledge, but for their creative participation" (p. 196). Just as the edges of the canvas constrain the artist, public hearing participants fill their three minutes with creative and evocative rhetoric that thematizes difference.

Rhetorical models of the public sphere emphasize the importance of connecting to specific audiences, creating common ground and presenting public arguments that draw from a range of rhetorical strategies. Iris Marion Young's (1996) "communicative model" of the public sphere "recognizes the cultural specificity of deliberative practices" (p. 123). By recognizing the need for publics to both address the cultural specificity of public participation and to serve in an informative capacity for citizens who are not familiar enough with an issue to be able to fully participate (p. 125), Young emphasizes the importance of rhetoric, mostly as it relates to a focus on audience and storytelling. Where Dewey sees rhetoric as a way of "[breaking] through the crust" (p. 183) of people's everyday lives, Young does not see rhetoric's role in publics as merely presentational. She employs rhetoric in an epistemic capacity. Rhetoric teaches by not only presenting information in engaging forms such as a narrative but by confronting people with difference. By being confronted with different perspectives we learn about our own perspective and those who oppose us (p. 128). Yet, place-based arguments contain the implicit notion of shared place despite ideological or political disagreement about whether or not to frack.

In response to those who hold that public discourse should be purely rational, we should ask what constitutes a public in the first place? What counts as public engagement? Publics are called into being when citizens engage with issues of common concern (Warner, 2010, p. 65). People can become members of publics in many ways, from simply paying attention to an issue to directly trying to impact that issue's outcome. Warner's perspective that the barrier for participation in publics is as low as simply paying attention is the view I support here. Every argument needs an audience, and those who pay attention are fulfilling the important role of auditors of persuasive public discourse. Still, the worry stands that when people engage with public issues in personal ways, they are acting more out of self-interest than what is best for all. Hauser (1999) and others who promote rhetorical models of the public sphere see self-interest as necessary due to the emotional connection public participants have to these issues that touch their lives (p. 51). However, this can be taken to an extreme when self-interested participants lose sight of the ultimate aim of their participation as the common good of all involved (Levasseur & Carlin, 2001).

At their core, those who advocate rhetorical models of the public sphere aim for inclusiveness by welcoming multiple modes of expression. By allowing and encouraging people to use discourse that does not reinforce the hegemony of the dominant in public spaces, they address Fraser's problem with the exclusionary nature of the rational-critical model. Karen Tracy (2010) sums up the rhetorical model when she says, "Democracy is about creating a space in which people can advocate passionately for what they believe" (p. 59). "Passionately" here invokes the inclusiveness of the rhetorical model, which invites the participation of those who bring self-interest and sometimes frame their arguments as pathos-laden appeals to both confront and be confronted by different perspectives.

When considering how people engage publics, it is common to assume that prior to entering the public sphere their minds are made up and their perspective on issues will not change. Models of the public sphere that encourage self-interested participation do not necessarily see a person's interests as a priori to the process of public engagement. There is much more value to public discourse when we see the development of interests as part of the process of public interaction (Fraser, 1990; Benhabib, 1992, 1996; Asen, 2003, 2004; Mansbridge, 1980). Rather than viewing such engagement as the expression of one's values and desires to impact policy, we can understand the deliberative process to be a "procedure for being informed" (Benhabib, 1996, p. 71). Benhabib sees this process as one in which people may enter a deliberative forum with "views and wishes," but many do not have a value hierarchy in place that translates to actionable agendas that can be considered by all (p. 71), which is especially a possibility in the setting of a public hearing. Rhetorical models of the public sphere, and even Habermas's model, which emphasizes discursive modes of engagement, engender the process of "discursive will formation" (Benhabib, 1992, p. 86). By participating, people learn about their values and develop the interests that guide deliberation. The value of the hearing is not only in the unilateral communication from citizen to power, but also the will formation that is built discursively and tied intimately to place. That is, it is a public good to be developed and circulated intertextually.

Just as researchers of public discourse should understand that participants develop interests and value hierarchies through the process of public participation, identity should be seen similarly. When people engage with a public issue, their identities should not be thought of as established and stable, but as subject to development and change through the process of public engagement. Especially in public settings where people are encouraged to use narrative and personal experience as evidence, personal and group identity are affected by the process. Habermas (1989) positions identity formation as private, prior to public participation, but others like Asen (2003) and Calhoun (1993) assert that it is socially constructed through the process of participation. Asen sees this as "collective imagining whereby people engage in a public process of identity formation" (p. 349). He states, "Active engagement of the collective imagination occurs in situations where advocates explicitly call upon their audiences to rethink relations to one another" (p. 351). Situations like this occur in public hearings through place-based appeals where participants challenge people's identification, as it is tied to the place that they live and the values assumed to be common among them.

At a Delaware Basin Commission hearing in 2018, two speakers who spoke within the span of fifteen minutes of each other invoked values and shaped the identities of those who wished to curb the expansion of hydraulic fracturing. One of them, Thomas Shepstone, author of the *Natural Gas Now* blog and self-described landowner, claimed, “You’re engaged in a massive land grab with the intent of making a wilderness of our land for the enjoyment of elite special interest funding the Delaware River Keeper.” Jeff Tittel, director of the New Jersey Sierra Club, responded, “I’m here representing our three million members and supporters nationally, but more importantly, our 100,000 property owners here in the Delaware Valley and the 17 million people that get their drinking water from the Delaware Valley. They have property rights, too, and what you do on your property could affect my property and my life and my family.” In two short speeches, these speakers call into question the meaning of owning property and how property rights are both defined and enacted. In particular, Shepstone addressed the “elite special interests” who he claimed were preserving wilderness for their own enjoyment, while Tittel, representing the special interest group of the Sierra Club, expanded their reach to all consumers of drinking water sourced in their area. These exchanges may not have swayed people from one side of the issue to the other, but they called into question the identities that people held and their defining values. They may prompt some to consider a more expansive definition of property rights and others to ponder if their desire to preserve wilderness for their enjoyment is a privilege of an economic elite. Exchanges like this one do not resolve conflict on their own, but may prompt people to consider how their position impacts others.

Public participation is built on the assumption that people really do share common ground and that they will to work together for the common good of all. Skeptics of this position, such as S. Michael Halloran, see as a myth the vision of community as a “unit or unity with shared norms to which members ‘belong’” (Simons, 2007, p. 61). Others, such as Patricia Roberts-Miller (2004), reject identification as a good of public participation. She claims, “The result of insisting upon identification as the central rhetorical strategy . . . is that public discourse tends to bounce between pandering and voices crying in the wilderness” (p. 111). Though I agree that as a “central” rhetorical strategy identification may divert the business of publics, it is the “wilderness” that brings people together to collectively act, crying or not. By this, I mean that in public hearings, citizens converge from seemingly isolated positions in the “wilderness” to turn the affective dimensions of public testimony into a collective value, which manifests as a common ground with opponents or through the solidification of perspectives on one side. Both will and identity formation are goods of the public process that the rhetorical model enables through its focus on inclusiveness.

WHAT DOES DEMOCRACY LOOK LIKE?

Normative models of the public sphere provide guidance to researchers who investigate publics. However, a competing view of democratic discourse has been developed by Karen Tracy (2010) that emphasizes the examination of existing democracy as a prerequisite to the application of theory. The model of "ordinary democracy" takes a discourse analytic perspective to the interpretation of public discourse, beginning with linguistic data and contextual features of publics that build up to theory. An ordinary democracy approach "[begins] with existing institutions and describes what is occurring in them" (Tracy, 2010, p. 4). In other words, we build our theory from the ground up.

Prior to Tracy's conception of "ordinary democracy," public sphere theorists such as Hauser (1999) advocated for rhetorical scholars to examine democracy in action (p. 31). To do this, researchers who take an ordinary democracy perspective mostly focus on democracy at the small scale, an approach that resonates with Dewey's assertion that we begin with the local: "We lie, as Emerson said, in the lap of an immense intelligence. But that broken intelligence is dormant and its communications are broken, inarticulate and faint until it possesses the local community as its medium" (Dewey, 1927, p. 219). As Dewey suggests, understanding rhetoric begins in the local manifestations of democracy. Tracy studies school board meetings in Boulder, Colorado. Others look at zoning hearings (Buttny & Williams, 2000) or city council meetings (Farkas, 2013). Foundational to Tracy's work is Jane Mansbridge's (1980) study of town hall meetings in a Vermont community where she takes an ethnographic approach to public participation, focused, in part, on language use. These studies of the language of democracy in its minor institutions offer deep insights into phenomena that are repeated in other sites of democracy, small and large. The ground-level focus on language in use in sites of ordinary democracy reveals not what normative models say people *should be doing* in democratic practice, but what people *actually do* when they participate.

These studies sometimes reveal common rhetorical strategies in public hearings, including the common place-based strategies that this book foregrounds. In terms of hydraulic fracturing, these strategies, or how people *do* democracy, manifest in one's connection to and concern for the land. While the rhetoric people use to express these connections and concerns binds them to a shared place, the different rhetoric they use to present their shared places reveals how ideology shapes and is shaped by their sense of place. As the focus on local and ordinary democracy brings issues of place-based identity, sense of place, and rhetoric of place to the foreground of democratic practice, what we may have seen as simply contextual becomes—for many who engage with the issue of fracking—front and center. Tracy and Haspel discuss the role of place in school board meetings as a tactic (2007, p. 156), a theme that is repeated in research on "place" in democracy, which Endres and Senda-Cook (2011) call "place-based argument." Such arguments "discursively invoke images or memories of place to support an argument, such as the summoning the melting of the arctic as a reason to stop global warming . . ." (p. 258). In other words, place-based arguments are discursive representations of place used for a rhetorical purpose. Note that though I am interested in how people imbue places with rhetorical power, I am not focused on how places themselves may be rhetorical, known as place-as-rhetoric (Endres & Senda-Cook, 2011, p. 258).

As stated, in democratic practice, places serve as common ground that may connect (or disconnect) people within a public. As Dewey (1927) reminds us, this is especially the case when publics operate on a local level: people have "difficulty in getting on either with or without [their] fellows, even in neighborhoods. [They are] not more successful in getting on with them when they act at great distances in ways invisible to [them]" (p. 131). Local democratic participation, as Dewey claims, does not make participation easy, but it does provide opportunity for the development of common ground that otherwise might not exist. When places in a community are utilized in argument as markers of common experience, the "invisibility" of neighbors is difficult to make disappear. Not only does a local focus implicate people in common action (despite ideological divisions), it provides frames of reference to ". . . work out relationships between the local setting and the larger frames of reference and obligation" (p. 242). Even in a time when local action may seem diminished due to the ease of digital communication, "the *worldwide does not abolish the local*" (LeFebvre, 1991, p. 86). In other words, the importance of the locally lived community does not disappear when issues such as land-based disputes, like hydraulic fracturing, impinge upon a community.

To understand how place is experienced, constructed and/o utilized as a rhetorical resource in such disputes, we must look deeply at places as physical topography and built environments as well as study the social relationships we find therein. Rather than viewing place as "backdrop," we should understand it as "the outcome . . . of social, cultural, political, and economic activity" (Powell, 2007, p. 2). LeFebvre (1991) claims that "(social) space is a (social) production" (p. 26), not a stable, immutable entity. Barbara Tversky (2003) explains the relationship between objective place, mental spaces and the rhetorical selection that makes place relevant to us:

> In human conceptions of space, the things in space are fundamental, and the qualitative spatial relations among them with respect to a reference frame form a scaffolding for mental spaces. Which elements or things are selected and which

> spatial relations are chosen as relevant depend on the space and the functions it serves us. (p. 67)

Tversky's focus on the function of mental representations of places fits into a rhetorical model of place. Rhetors make places relevant to their audience by selecting elements to emphasize based on their usefulness. For instance, in hearings on Marcellus Shale, a speaker may represent a pastoral scene that emphasizes the orderly construction of a well site by focusing on structural elements of the scene like paved roads. From LeFebvre's perspective, these "representations of place" are society's signs of how socially constructed places are made relevant (p. 26). Researchers in environmental psychology go so far as to say, "descriptions and evaluations of place are arbitrary and can be contested to the extent that they are tied to people's projects at hand" (Auburn & Barnes, 2006, p. 44). This perspective, though quite radical, assigns meaning to place only through its rhetorical purpose. Others, like J. Nicholas Entrikin (1991), take more of a geographer's perspective, seeing a contest between two modes of thinking about place; one views place as objective reality and the other as subjective and culturally entangled. Through his narrative approach, termed "emplotment," Entrikin describes how the objective representations of place are made meaningful through causal connections with subjective agents much like the way narratives employ plot to make setting and characters significant. Similarly, Michel de Certeau (1984) claims that narrative transforms places. Storytelling "tell[s] us what one can do in [a place] and make out of it" (p. 122). As a tactic used in the site of a public hearing, narrative shapes perceptions of place and allows participants to contest the meaning of place as it is relevant to the local setting and beyond.

As a framework for our thoughts or as narrative connections between agents and their setting, place also serves the rhetorical canon of memory. The "memory palace," as it is known, has been used since ancient times by locating memories in imagined places. An example of its function is found in the *Rhetorica ad Herennium*, which recounts the memorable story of Simonides who, after the collapse of a roof onto a hall full of people, was able to recall the positions of those people in the room so as to give them a proper burial. By mentally "walking" through place, items or people can be easily recalled. The efficacy of human connection between place and memory has developed through our evolution as hunter gatherers; through our brain's hippocampus, we are physically wired to store spatial memories (Jeffrey, 2018). In a very real sense, place and memory are inseparable (Kalin & Frith, 2016, p. 232). This is important when considering the rhetoric of place because through language, we shape what is remembered, recontextualized, and circulated through public discourse.

When we study how people engage with actual, physical places and how these sites manifest in public argument, rhetorical scholars and those interested in place-bound public issues are afforded a lens through which to understand rhetoric, place, and public controversy. We must understand that sharing a *sense of place* is a powerful component to the human experience. As Douglas Reichert Powell (2007) explains, senses of place are also discursive constructions; they are not fixed, but

> . . . ongoing debates and discourses that coalesce around particular geographical spaces. Furthermore, it is by looking at those features of a place that seem, at least superficially, to be the permanent stable markers of its identity that we can begin to see the dynamic, evolving, and rhetorical qualities that create and sustain what has often been taken (reductively) to be an ineffable or ethereal, sensory property: the "sense of place." (13)

When we consider the features of the built environment that citizens represent in their public rhetoric, interpreting their configuration of place, we can learn far more than just what gives a place a particular feel. This analytical process allows us to comprehend the nature of public rhetoric.

Chapter 2

Vast Pennsylvania, Unique Pennsylvania

Synecdoche and the Representation of Place

Perspectives on environmental protection, energy politics, and economics collide in local venues near sites of natural gas extraction. Despite the very real presence of this industry, even those who live near natural gas extraction operations rely on abstractions to make sense of the process and its impact. The frequent use of abstractions, however, hinders understanding of and communication about the risks of energy development among stakeholders. Participants in public hearings, representing key stakeholder groups in this issue, tend to make sense of risks by analyzing the impact of fracking on the concrete places in their communities. In the process, place becomes an indicator of impact as well as a resource of deliberative argument, providing a concrete referent for claims about environmental protection, energy politics and economics.

In this chapter, my aim is to broaden our understanding of place-based arguments by examining each of the 886 place references in one public hearing on hydraulic fracturing and to offer researchers a coding scheme to study the rhetorics of place in environmental controversies. I situate this study in the conversations on critical regionalism that seek to disrupt the flattening of local difference. Although people disrupt regions with rhetoric in ordinary democracy, I show that they also help build them. Place-based arguments, as described by Endres and Senda-Cook (2011), are simply descriptions of places or the invocation of memories tied to places (p. 258). By examining the rhetorical purpose of each place reference, I show that the value of place in deliberative argument is much more than to provide a location or setting for the speaker's comments. I also seek to understand how place-based arguments are used differently by subsets of participants. To guide this investigation, I pose three main questions:

1. *What is the discursive form and function of place in this public hearing?* What rhetorical choices do people make as they invoke places in their speeches? What rhetorical impact do they intend these place invocations to have on their audiences?
2. *Are places represented differently by any subsets of the hearing? If so, how are places differently represented?* Do supporters of hydraulic fracturing depict places differently than those who oppose fracking? Do other groupings of people (e.g., farmers, landowners, environmental groups, sportspeople) reveal regular patterns of place depiction?
3. *What is the deliberative function of place-based appeals?*

I seek to answer these questions by analyzing a March 10, 2011, Allegheny County, Pennsylvania public hearing. This analysis reveals six main classes of place in public hearing discourse: natural, industrial, commercial, educational, political and home places. These classes of place are used by speakers representing all sides of this issue. Of these six, three categories—*natural places*, *industrial places*, and *home places*—help me answer questions two and three and are therefore the focus of this chapter. Each of these categories of place is constructed differently by subsets of hearing participants to help achieve their rhetorical aims.

The most striking division of this public hearing is between those who support and those who oppose hydraulic fracturing. I find a distinct difference between the ways that pro- and anti-frackers use place-based arguments. For instance, the representation Dr. Judith Kaufman, an anti-fracker, makes of industrial sites differs from that of pro-fracker Ed Valentas. Kaufman, a doctor and Allegheny County resident, compares television commercials with the reality of living in this place:

> (1) . . . the ads by the industry present this idyllic picture of the farmers and landowners who have nothing but love for their Range Resources but never show the actual fracking ponds, condensation tanks, ravaged access roads, 120-feet drilling rigs, flares and man camps that become the next-door neighbors.

Kaufman references both industrial and home places, situating the apparatuses of industry in close proximity to home. Alternatively, Valentas, a pro-fracker, presents industrial activity from his personal experience and emphasizes the vast extent of this activity without getting into specifics:

> (2) I've drilled and fracked, operated about 1,000 wells in six states. And some of these were twice as deep as the Marcellus Shale, and some were horizontal. . . . But they were all done, all fracked, and without incident. People are acting

like this fracking is new, and to a lot of these people, it is. It's like a UFO coming into their neighborhood.

While Valentas creates the impression that any opposition to fracking must be due to a failure to comprehend the process, Kaufman assumes that industrial machinery and its workforce would not be welcome in any audience member's neighborhood. In spite of their thematic difference, these two arguments, and the subsets of which they are a part, construct place in distinct yet predictable ways. Despite lacking formal coordination prior to giving their speeches, pro- and anti-frackers use similar place-based rhetoric to advance their claims. This provides an opportunity for rhetorical scholars to better understand how ideology shapes place and, reflexively, how affiliation with places can shape a person's ideology. Those seeking to engage with this or other place-based disputes can expect similar rhetoric, which presents an opportunity to either amplify or oppose these appeals.

The primary difference between the discourse of those who support and resist fracking is the proximity imagined between a speaker, their audience and the subject of their discourse.[1] I have found that pro-frackers tend to present places from afar, while anti-frackers present places up close. Representing places from varying distances facilitates the representation of the pro-fracking stance that a geographic area is protected and the anti-fracking stance that it is threatened. Pro-frackers further build their argument by discussing "place" on a large scale, as a large whole made of many parts, while anti-frackers focus on the parts as individually valuable and threatened. In an anti-fracking discourse, a threat to a part is a threat to the whole. The place-based arguments of both stances reach beyond the scope of the hearing itself, which is meant to be confined to Allegheny County. Understanding the way that place is framed allows for an understanding of the way citizens conceptualize their community's role in the larger currents of society as well as the relationship between place and personal identity.

The focus on parts and wholes is a common feature of place-based rhetoric for this issue and is known by rhetoricians as synecdoche. Speakers use synecdoche whenever they represent places as either individual parts of a greater whole or a whole that subsumes its parts. As mentioned, anti-frackers tend to focus on how the part implicates the whole while pro-frackers tend to focus on the whole, implying that the parts are the same. The presentation of the synecdochic relationship between the whole and its parts allows rhetors to make claims about places they do not explicitly represent, but that are inherently connected to the places they do represent. That is, it allows speakers to build relationships between their individual experience in a place and that of others who know another part of the same whole but do not share knowledge of the same place. Synecdoche allows specific place references to connect to other areas of concern. In the context of public rhetoric, people use synecdoche to relate their self-interest to the broader interests of their community or to situate their community in a larger context, which is done when we consider the role of a county in Pennsylvania in US energy independence.

When pro-frackers use synecdoche to construct region-level depictions of place, I have found, they tend to downplay local difference. At the same time, anti-frackers emphasize local difference by stressing the unique importance of specific parts of places and then extending this unique status to other parts within the whole of the Marcellus Shale region. These different place-representation strategies are important because they show both how pro- and anti-frackers communicate the value of their place and how they engage in a process of "region-making." That is, synecdoche helps to provide a rhetorical angle on critical regionalism. Beginning in the fields of architecture and geography, this movement encourages people to emphasize local difference to resist the loss of regional diversity in a globalizing world (Powell, 2007). These regionalizing forces are mostly seen as corporate entities that prioritize identical fast food restaurants, big box stores, and other capitalistic ventures over local difference. The analysis in this chapter shows that though some speakers work to represent the uniqueness of their own place, many find power in region-making. Regionalizing discourse facilitated their engagement with a complex technical issue like hydraulic fracturing. Although regionalism may obscure local difference, it may also provide fortitude to citizens who are engaging with their democracy.

SYNECDOCHE: THE PARTS AND THE WHOLE

Though synecdoche is commonly used to reference a part representing the whole, Burke (1969) first defines synecdoche as "representation" before clarifying that it is "part for the whole, whole for the part, container for the contained, sign for the thing signified, material for the thing made . . . , cause for effect, effect for cause, genus for species, species for genus, etc" (p. 503). By offering the more specific explanation of synecdoche after the broader definition, Burke implies that synecdoche is not limited to a part representing the whole or a part standing in for a whole, as in metonymy. He explains, "All such conversions imply an integral relationship, a relationship of controvertibility, between the two terms" (pp. 507–508). This "controvertibility" allows us to understand synecdoche as the act representing one entity while implying the same characteristics for a synecdochically related other. Simply put, synecdoche can describe both relationships of part-to-whole or whole-to-parts.

In the discourse of pro- and anti-frackers, the two types of synecdoche emphasize different ends of the part-whole relationship: anti-frackers tend to stress the "part," pro-frackers the "whole." Perelman and Olbrechts-Tyteca (1969) write, "Usually the relation of the whole to its parts is dealt with quantitatively: the whole includes the part and is consequently more important" (p. 231). Though it may be that the whole is considered more important than the parts, by focusing on the parts rhetors highlight the small-scale and distinct instances of nature (e.g., parks, state forests) that are parts of the whole in Pennsylvania, an important strategy in the issue of hydraulic fracturing. Explaining the uniqueness of each part increases the value of each individual instance of nature while supporting the claim that the whole is made of many unique parts. Meanwhile, pro-frackers deny the uniqueness of parts and instead make claims about the whole. As both sides show, the rhetorical function of synecdoche that emerges in the hearings advances a controvertible relationship between the whole and its parts, which may be utilized toward different ends in a deliberative setting.

In this manner, speakers regularly use discourse strategies to represent places as either part of a larger whole or as a larger whole that subsumes its parts. This is the way synecdoche works as a controvertible concept. As shown, each form of synechdoche is used primarily by one group. The different modes of representing place by pro- and anti-frackers do more than help speakers support their position on the issue; they actively work to shape "undifferentiated space" into meaningful place, a process of familiarization. That is, when the space is well enough known, we may "endow it with value" (Tuan, 1977, p. 6).

The use of synecdoche as an explanatory concept in an environmental controversy has precedent (Moore, 1993), though its application to hydraulic fracturing and the places where it occurs is new. In his noted work on the spotted owl controversy, Moore demonstrates how synecdoche can obscure political debate, guiding the experience of the controversy for competing groups. It thereby functions as an "organizing principle" for debate. Moore explains, "As the owl guides the conflict, it also obscures the larger issue of preservation versus extinction, to create a separate reality for debate" (p. 259). Not only can synecdoche function to turn attention away from larger issues, but, Moore explains, it creates a "contrived conflict" that replaces the actual issue at stake and provides no path toward resolution. Moore's approach to synecdoche demonstrates that the part may not only obscure the whole, it may make participants feel they are discussing the same issue when, in fact, the part (owl) at the center of the controversy is part of a different whole ("owl as indicator" for environmentalists and "owl as scapegoat" for loggers). Like the spotted owl controversy, hydraulic fracturing discourse features synecdoche as an organizing principle that obscures debate. However, there is no singular entity like the spotted owl on which to fix one's attention. Instead, groups compete to represent the part-to-whole or whole-to-parts of the many places invoked in this hearing.

Critical Regionalism and Regional Rhetoric

Critical regionalists push against the synecdochic representation of large geographic areas in ways that subsume smaller parts and advocate for more mindful construction of regions as a critical enterprise. They uphold that small-scale differences are important to maintain and the uniqueness of any one part should not be swallowed by the whole. In this manner, Douglas Reichert Powell (2007) reminds us that the "Marcellus Shale region" is not a stable political or geographic entity. Regions are defined loosely and their boundaries may change depending on context. Andrew Wood (2012) explains this point when he says, "A place is a fact; a region is a choice, and not necessarily one made by its inhabitants" (p. 290). Synecdoche, on the other hand, suggests that characteristics of the whole belong equally to the parts, as exemplified in the following Texas-based newspaper report: "For perspective, if the Marcellus Shale region were a country, its natural gas production would rank third in the world, after Russia and the rest of the U.S" (Begos, 2013).

The delineation of the Marcellus Shale region in the above citation echoes Jenny Rice's (2012) argument that region-making is inherently rhetorical. She explains that "regional rhetorics provide alternative ways of framing our relationships and modes of belonging" as well as "[disrupting] given narratives of belonging that are framed on a national level and between individuals" (p. 203). Rice and her colleagues who participated in a special issue of *Rhetoric Society Quarterly* on the topic of regional rhetoric (Tell, 2012; Olson, 2012; Clark, 2012; Green & Kuswa, 2012; Wood, 2012), view regional rhetoric as a tool that identifies people with each other, resulting in the building of a public; as distant people are linked together in place-related affiliation, they gain power. As a common sense of place is created, it influences the performance of individuals who identify as members of a particular public (Wood, 2012, p. 291).

The work of regional rhetoricians and critical regionalists allows us to better understand how attendees (and those members of the public not present) at the Allegheny County hearing create and experience a sense of place. Though most of these participants do not specifically refer to region, their differing representations of place work to construct regional understanding and affiliation. Although the role of synecdoche in this process has been overlooked by regional rhetoricians, it provides an important addendum that can show us how people reach beyond their local contexts as they strive to shape their world.

AN ANALYTICAL FRAMEWORK OF PLACE IN PUBLIC DISCOURSE

The public hearing on March 10, 2011, drew the hydraulic fracturing industry, grassroots environmental groups, the media, and many concerned citizens to the Allegheny County Courthouse to share their perspectives on the future of gas extraction in their county. As video evidence reveals, the council chambers were so packed with people that many participants and audience members needed to remain outside until speakers exited (Jansa, 2011). During the meeting, 75 speakers gave testimony, most of whom announced themselves as residents of Allegheny County. Gas industry workers, scientists, politicians, healthcare workers, environmental advocacy group members and other citizens spoke on the broad topic of Marcellus Shale. Divided by ideological lines, 49 speakers were against hydraulic fracturing while 26 supported the expansion of this technology. Chair Charles J. Martoni, one of 11 council members present (and one council member who telephoned in), opened the meeting by addressing the large crowd size:

> (3) We appreciate you coming up. And we do have people waiting. We'd appreciate, if you're done, to leave the room for another person to come in that might be in a hallway. We'd appreciate that. Okay? Under the Fire Code, we're only allowed so many people in here. We can't just stack the place. We have to be careful we don't break the rules.

Indeed, the issue of hydraulic fracturing in Allegheny County drew crowds and the testimony ran past midnight. Citizens from Allegheny County and beyond met together to argue for the fate of their county, commonwealth, and country.

The hearing was one of many held not only around the Commonwealth of Pennsylvania, but across the United States. It is representative of the others in its proceedings as a noninteractive hearing where citizens had three minutes to speak before a local legislative council and other participants. Although it was dominated by the opposition, there was a strong showing of speakers with industry relationships through coalitions and business partnerships who spoke of jobs, health, families, science, relationships, and the places in their community. Drawing from their experiences, they connected their know-how with the larger influence of national and international energy development as well as personal concerns over healthy and safety. In as far as the people represent ordinary people; the hearing becomes an instance of what Karen Tracy describes as ordinary democracy. Nonetheless, the systematic rhetorical analysis of the Sargent's Court Reporter's transcript of this hearing reveals the extraordinary nature of the public rhetoric of place used.

The analysis here is not specific to this corpus or this case, but can be used by others to examine place-based rhetoric in their own research sites to reveal how democracy works in its ordinary institutions. Using a process of analytic induction through multiple readings of each speech, categorizing the typologies of place-focused language use, and determining intended effect, I offer a framework of place-based rhetoric. What I present builds on work by discourse analysts and sociolinguists seeking to explain the role of place in discourse (Schegloff, 1972; Entrikin, 1991; Chilton, 2005; Myers, 2006; Van Leeuwen, 2008). This research makes clear that place is flexible, as "any place can be described in broader or narrow terms" (Myers, 2006, p. 322). When rhetors represent places, they select among the many options available to them from geographical to relational formulations (Schegloff, 1972). My own coding process made apparent the importance of "place" as a multifaceted concept in this hearing. As Myers (2006) cautions, if we pass over place-naming as a conversational practice light on meaning, we miss the deep connections between place formulation and identity.

To produce the findings of this chapter and this book more broadly, I employed a systematic and data-driven empirical method that was responsive to movements in the study of rhetoric and composition (Barton, 2000), while working to retain the complexity of the language under study (Geisler, 2018). The first step in this process was identifying all references to "concrete places." Concrete places are identified by proper nouns that name a geographic, topographic, or structural entity no matter the size. This may range from a mention of "Pennsylvania" to "North Park" to "Stevens Creek." Identification of each instance of a concrete place in the hearing led me to the realization that place is also constructed through general place references such as "the state," "parks," or "streams." When speakers reference concrete or general places, they can indicate assumptions about their audience regarding co-presence or shared knowledge of place that is important to capture (Schegloff, 1972, p. 84). The choices to represent places concretely or generally also may position the speaker in varying degrees of proximity to the place itself (Chilton, 2005, p. 88).

Each instance of a place reference, concrete or general, was cataloged, resulting in 886 place references. In addition to place references, I noted the speaker's name, their stated residence, their side of the issue (pro or anti), professional affiliations, and the argumentative purpose of the place mention. Determining the argumentative purpose of each place mention resulted in a more systematic coding scheme that focused on the class of places, the speaker's relative proximity to the place, and their evaluation of those places.[2]

Classes of Place

My coding scheme first examines classes of place and reveals six overarching categories. The first class is (1) *Natural places*, which include any geological or topographic entities expressed as noun phrases. These are most often streams, rivers, parks, forests, and woodlands. Natural places are also more broadly defined when speakers refer to an entire county or state as predominantly natural. For instance, if a speaker were to state, "Pennsylvania's network of pristine streams and rivers . . . ," I coded "Pennsylvania" as a natural place.

The second class of places is (2) *industrial places*, which consist of well sites, chemical plants, water treatment facilities, and other industrial sites and apparatuses. Like natural places, I counted references to geopolitical entities as predominantly industrial as industrial places. Some speakers refer to Pennsylvania as a state that has seen heavy resource extraction for more than a century, which I also count in the industrial class. Admittedly, these speakers are not claiming that Pennsylvania is *only* industrial, but by constructing the state in this way they shape how their audience perceives it.

The three following classes of places—(3) *political places*, (4) *commercial places*, and (5) *educational places*—were coded in much the same way as the previous two. Noun phrases that referred to places for their political activity (e.g., referring to Harrisburg, Pennsylvania's capital, as a metonym for the state government) were coded as political. Places that are predominantly used for commercial activities, such as references to shopping centers or other businesses, were coded as commercial. Schools and universities were also mentioned by some speakers, especially in reference to well-site setback regulations; these were coded as educational. These three categories did not produce significant insights about the data and will not be discussed further in this chapter.

The final coding category for classes of place, (6) *home places*, was divided into three sub-categories. Home places were invoked as an individual's home, as the plural form "homes," or as a geo-spatial entity that serves as home (e.g., "Pennsylvania has been my home for ten years").

Table 2.1 Classes of Place Framework

1. Natural—Noun phrases involving descriptions of specific or general places that are noted predominantly for elements not developed by humans and may include geopolitical areas characterized by natural features
2. Industrial—Noun phrases involving descriptions of specific or general places predominantly noted for elements related to the development of resources and/or goods that may include geopolitical areas characterized by industrial features
3. Political—Noun phrases involving descriptions of specific or general places marked by political boundaries or designations
4. Commercial—Noun phrases involving descriptions of specific or general places marked by activity related to business, sales, marketing, economics, or other non-industrial enterprises
5. Educational—Noun phrases involving descriptions of specific or general places related to schools, universities, museums, or other spaces noted as sites of learning
6. Home—Noun phrases involving descriptions of specific or general places of domicile ranging in scale from personal dwellings, to one's county, state, or country

Beyond these six overarching categories, I also coded for other uses of place references such as the procedural constraint of giving one's home address at the beginning of a speech, the use of general place references to advance an explicit evaluation of places as "threatening" or "threatened," "beneficiaries" or "protected." For instance, I coded an excerpt from one speaker who describes "a huge parcel of land owned by a lady from Florida" that has been leased to the gas industry as a "threatening place," not as one of the six larger classes, because rather than emphasizing the place itself, the description "threatening" referred to the status of its ownership. This accounts for the difference between the 886 initial place references and the 543 that fit into the 6 classes I discuss in this chapter. These other categories have helped me understand the rhetorical purpose of concrete place references.

Distant/Proximate Places

In addition to noting the class of each place reference, I also considered whether the place was distant or proximate to the hearing and whether it was referenced to provide a positive or negative example. To make a determination regarding proximity, I draw from Piotr Cap's (2013) spatial proximization framework, which considers elements inside or outside of the deictic center of a place reference (numbers 1 and 2 in table 2.3 are from Cap's framework for spatial proximization). For this hearing, the deictic center is most clearly seen as Allegheny County, though this may vary based on how speakers characterize what is nearby. This relative positioning of the deictic center is important. When people process language they orient characters and features of the discourse relative to themselves (Chilton, 2004, p. 57; Cap, 2018, p. 384). Typically, entities that are closer to the deictic center are evaluated positively and distant entities are evaluated negatively (Cap, 2018),

Table 2.2 Classifications for Place References in the March 10, 2011, Allegheny County Public Hearing

Class of Place	*Number of Uses (out of 543 total uses)*	Examples
Natural	113	*Stevens Creek; Allegheny National Forest; our rivers*
Industrial	135	*Well site; frack pond; four shallow wells*
Political	155	*County facilities; state lands; City of Pittsburgh*
Commercial	38	*Stores; restaurants; 20 farms in this area*
Educational	15	*Neighborhood K-to-eight public school*
Home	87	*Our home; Allegheny County; Pennsylvania*

but in the case of environmental issues like hydraulic fracturing, this is not always the case.

In this hearing, if the reference was within Allegheny County, I considered it a proximate place. If the reference was outside of Pennsylvania, I considered it a distant reference. For references that were within the state of Pennsylvania but outside of Allegheny County, I relied on a secondary set of questions that asked if the speaker signaled a close or distant relationship to Allegheny County. For instance, if a speaker listed several counties including Allegheny (e.g., *We have to consider the air quality issues we'll begin to experience in Greene, Venango and Allegheny counties*), I consider all of the listed places as proximate and within the deictic center. If the place was referenced to create a distinction with Allegheny County, I considered that distant, as seen through the mention of Erie County in the following example (e.g., *While northern counties like Erie don't have as much to worry about, Allegheny County is right in the epicenter)*.

Not all of the 886 references to place initially identified create a sense of proximity or distance and not all provide a clear positive or negative evaluation of these places. Many general references to places do not specifically locate the entity they describe and do not evaluate it. For instance, when speakers generally referred to "Marcellus drilling sites" or the apparatuses of fracking like "condensation tanks," I did not code them as evaluating proximate or distant places.

Evaluation of Place

Five additional codes were used to illuminate the evaluative stances of speakers toward places. Speakers not only invoked places in their public discourse, they also evaluated these places to emphasize their functionality, connect them to certain actions, and position them in value hierarchies (Van Leeuwen, 2008, p. 101). The evaluations prominent in my data are primarily hierarchical, as speakers relayed the symbolic importance of these places to their communities. These are: (11) *place as beneficiary*, (12) *threatened place*, (13) *threatening place*, (14) *protected place*, and (15) *shared place.* Code (11) was used when a speaker referred to positive change experienced in a place, usually

Table 2.3 Proximity of Place Framework

1. Proximate Place—Noun phrases construed as elements of the deictic center
2. Distant Place—Noun phrases construed as elements outside of the deictic center
3. Positive example—Element inside the deictic center used to support a favorable claim
4. Negative example—Element outside of the deictic center used to support an unfavorable claim

Table 2.4 Places Were Coded as Distant/Proximate and Positive/Negative

Evaluating Proximate/Distant Places	*Number of Uses (out of 239 total uses)*
Distant place (positive example)	35
Distant place (negative example)	60
Proximate place (positive example)	85
Proximate place (negative example)	59

Table 2.5 Evaluation of Place Framework

1. Place as beneficiary—Verb phrases that characterize a place as recipient of a positive good
2. Threatened place—Verb phrases that characterize a place as the potential or continual recipient of negative impact
3. Protected place—Verb phrases that characterize a place as removed or shielded from negative impact seen elsewhere
4. Shared place—Verb phrases that characterize a place as owned or experienced jointly among community members

associated with financial gain. Places were considered threatened when speakers made claims that a place would be negatively affected in the future. This was used often by anti-frackers to create the sense that natural places were threatened, and by pro-frackers to invoke economic threat. In general, *threatening places* was used when a place was positioned as a negative influence on its surroundings, while *protected place* was used when the argumentative purpose of introducing a place was to state that it was in no danger. *Shared places* are those that speakers characterized as of common concern to all audience members.

As above, often general place references did not clearly evaluate a place and therefore did not receive a code for this category. This is why 443 of the initial 886 place references are coded in this category.

FINDINGS: ANTI- AND PRO-FRACKER STRATEGIES FOR REPRESENTING PLACE

As we have seen, speakers on both sides invoke place, though the analysis of their place-based appeals reveals marked differences between how pro- and anti-frackers represent places. By classifying natural, industrial, and home places, I explain the prominent sites of place-based appeals and differentiate how these sites are operationalized in public rhetoric. The primary difference is the scale and specificity of place references.

Table 2.6 Codes and Frequency of Codes Used to Describe Evaluations of Places

Evaluation of Place	*Number of Uses (out of 433 total uses)*
Place as beneficiary	84
Threatened place	165
Threatening place	72
Protected place	19
Shared place	93

Natural Places

Because hydraulic fracturing is typically done in rural areas, many participants in the Allegheny County public hearing spent considerable time shaping public perceptions of nature. Out of the 75 participants in the hearing, 44 make at least 1 natural place reference. Out of those 44, there are 113 references to natural places. "Natural places" here refers to waterways, forests, parks (local, state, and national), and larger geopolitical entities referenced for their natural qualities. For instance, when a speaker refers to "rural Wyoming" in relation to the expectation that the natural environment there should be pristine, they make a natural place reference. In this section, I report on findings regarding the type of natural places referenced and explain how natural places are discursively constructed and used rhetorically by anti- and pro-frackers. As mentioned, I find that pro- and anti-frackers represent nature in distinct ways through synecdoche, the former emphasizing that natural places are vast, resilient and plentiful while the latter stressing that they are unique and shared by all community members.

Water

Because water is essential to hydro-fracking, it was the most contested "place" in this hearing. In the ideal, safest hydro-fracking process, water is drawn from rivers and streams, combined with proprietary chemical mixtures, shot deep down into the earth to fracture shale, drawn out and expelled into collection ponds, and then transported and treated. Unfortunately, this is not always what happens; anti-frackers reject the notion that the process is so cut and dried when they refer to past contamination events and speculate about the likelihood of spills. At the same time, speakers discuss water as a source of recreation and local pride. Out of 113 total references to natural places, 47 of these deal with waterways in some form, most mentioning the Monongahela and Allegheny Rivers, the 2 rivers that converge in Pittsburgh.

Parks

Due to proposals to allow fracking on public lands, state and county parks and forests are at stake in this hearing and referred to by pro- and anti-frackers alike. While one side claims that county parks are dilapidated and need the financial benefit fracking would bring, the other emphasizes the status of these places as sites of recreation, not industrial activity. Though parks are invoked only 16 times by name, speakers spend considerable time discussing their ideal plans for them and what they mean to their communities.

Counties and States as Natural Places

In 13 instances, speakers refer to geopolitical entities as natural places. Most commonly, pro-frackers refer to particular townships within Allegheny County as rural enough to be conducive to hydraulic fracturing. For instance, Patrick Imbrogno describes Moon Township as "very rural living," distinguishing it from the more urban areas, notably Pittsburgh, where many anti-frackers in attendance reside. The Commonwealth of Pennsylvania is also characterized as a natural place in comments by Mark Sommers, who states:

> (4) I deeply value and appreciate Pennsylvania's vast natural resources and natural beauty. The legacy that I leave for my children and future generations will be that of environmental stewardship and energy independence.

As Sommers shows, natural places can be characterized as a "resource" as well as a site of "beauty" and recreation. States distant from Pennsylvania are also referenced for their common association with the natural environment, especially Wyoming and Colorado.

Natural Places

Comparative analysis reveals notable differences between how pro- and anti-frackers construct natural places and use them for rhetorical effects. Their rhetoric reveals contrasting views of the value of natural places. Anti-frackers, analyzed in the first section, emphasize the uniqueness of natural places and their status as a shared public resource.

Anti-Frackers: Unique and Shared Nature

By emphasizing the unique quality of a natural place, speakers assign a privileged status that requires special care. The last of anything must be protected; a truly singular object should be highly valued. Anti-frackers imply that the places they reference are unique enough to hold this value. Their focus on the uniqueness of specific sites comes forth in the use of

Figure 2.1 McConnell's Mill State Park, Lawrence County, Pennsylvania. *Source*: Photograph by Michael Yatsko.

synecdoche, through which they raise the status of individual places as important parts of the larger whole. At times, they specifically invoke the uniqueness of a place, as exemplified in the comments of Erika Staaf of PennEnvironment:

> (5) Allegheny County's parks are true gems and are special places that bring communities together. . . . They add an unquantifiable value to the county and to the region.

Though Staaf refers to multiple county parks, she ascribes them a value that relies on their unique (or at the very least, rare) status within the community as "gems" and "special." This special status, Staff argues, should afford these places protection, which can only come by protecting the entire county.

Anti-frackers also create a sense of uniqueness to natural places by referring to them in concrete terms. Perelman and Olbrechts-Tyteca (1969) explain that "there is a close connection between the value attached to what is concrete and what is unique: by displaying the unique character of something we automatically increase its value" (p. 77). Though Perelman and Olbrechts-Tyteca describe "concrete values," their explanation insinuates that they are the same as concrete entities: "A concrete value is one attaching to a living being, a specific group, or a particular object, considered as a unique entity." In practice, this sense of concreteness can be created simply by referring to an entity by its proper name.

The use of proper nouns is one major distinguishing feature of anti- and pro-fracker discourse. Anti-frackers refer to natural places by their proper name more often than do pro-frackers at a rate of 59–42 percent. The most common proper nouns used by anti-frackers are derivations of the Monongahela River (e.g., the Mon) and the Allegheny River. These waterways, which are close to the site of the hearing, are continuously drawn on to characterize the stakes of less-than-desirable outcomes of the fracking industry.

The idea of uniqueness constructed by anti-frackers relies on a confined view of places in relatively close proximity to the hearing. They do not argue that Allegheny County holds nature that is unique to the world so much as they emphasize the uniqueness of nature within the smaller scale of their community. This argument is one way that anti-frackers establish risk, stressing the value of natural entities in the role they play in local lives. For instance, city councilman Doug Shields testifies:

> (6) I have talked to scientists from Carnegie Mellon and elsewhere, we're now getting bromides in the river, in the Allegheny River where the City of Pittsburgh intakes are.

By connecting the concrete natural entity, the Allegheny River, with city water intakes, Shields discursively links a topographical feature to a unique relationship with the residents of Pittsburgh.

Anti-frackers like Shields make explicit moves to connect community members to risk through specific place references. By pinpointing the Allegheny River as the unique source of water for the city, he establishes that a threat to the river affects community members. The argument that risk is shared is also created through anti-frackers' frequent use of the plural possessive pronoun, "our," with general reference to natural places. Joni Raboniwitz, an anti-fracker and self-described activist, demonstrates this use of "our" in her discussion of a farming conference she attended:

> (7) In this group, there's a growing opposition to Marcellus Shale drilling across the whole state . . . as people who grow the food we eat and keep *our streams* clean [italics added].

Rabinowitz provides evidence that not all anti-frackers focus on place up close. She relies on the synecdochic structure typical of pro-frackers, which foregrounds the whole to make claims about the parts. At the same time, by qualifying the streams as "*our* streams," she implicitly invokes individual streams with which the audience is familiar. This can be seen as implicit part-to-whole synecdoche that draws on the unique to make claims about the whole.

This use of the possessive pronoun "our" can also be seen as an involvement strategy that asks audience members to imagine the place reference from their own perspective. Similarly, Kevin Loughney, another anti-fracker, describes first-hand experience at a drilling site:

> (8) I was awed, sickened and sad to see these things going on in highly populated areas, state lands and *our forests* [italics added].

This same use of "our" in reference to a general natural place may also invoke a connection between the audience and specific forests they have experienced. Loughney differs from Rabinowitz in the speech he gave before Rabinowitz made the above claim, where he provides a detailed account of a specific frack site:

> (9) The frack that I saw was aided by tanker after tanker of liquid nitrogen. They said it was a small well and that it was cleaner than using more fracking fluid. Liquid nitrogen into our ground, directly into our ground.

Loughney's example of the use of synecdoche is typical of anti-frackers who construct a scene that is meant to be a vivid part of a whole that is under the same threat. His use of parallelism in "our ground" and "our forests" explicitly creates the expansion of part-to-whole for the audience. They claim that each part is "ours" and that if the parts closer to the speaker are treated this way, then the parts closer to the audience will be too.

In summary, anti-frackers construct natural places as either unique entities that are at risk and should be valued for their unique status or they construct risk as widely felt by all community members via general natural place references preceded by the possessive "our." Synecdoche is typically used to create a sense of risk from specific parts to the greater whole. This strategy helps speakers create a sense of community risk not only to individual parts that may be unfamiliar to the audience, but to those parts they can recall that share characteristics of what has been described.

Pro-Frackers: Vast and Resilient Nature

Pro-frackers emphasize the vastness of nature by using general more often than specific place references along with plural nouns and descriptors of size. Supporters of fracking also reinforce the notion that nature is resilient and will be able to withstand any inadvertent pollution. By referencing quantifiable evidence such as Herman Edwards's claim that "the county has thousands of acres of parks," they create an image of nature as vast. Patrick Imbrogno, cited earlier, also argues for the vastness of nature:

> (10) I live in Moon Township, very rural living. It's a lot different from where you people are talking about living. But there are a lot of rural areas here, and there's a lot of areas . . . .

By focusing on the vastness of the many rural areas, pro-fracking speakers use synecdoche to emphasize a whole that subsumes its parts. The implication is that if some land is despoiled, there will be plenty more left unscathed.

Other pro-fracking speakers refer to "nature" on a global scale, directly challenging the local view of anti-frackers. Adrian Markocic cautions against this perspective when he claims:

> (11) Every time you use your iPhone, you drive your Prius to Wholefoods or your farmer's market, you're contributing to the destruction of the Niger River Basin, the Amazon Delta and Middle East deserts.

By invoking distant natural places affected by the oil industry, Markocic attempts to expand the working definition of "nature" in this hearing. Most anti-frackers only discuss nature that is in close proximity to the hearing, at least within the state. Markocic, a pro-fracker, reminds the audience that all nature is to be valued; only protecting nature at home translates into "selfishness, prevailing not-in-my-backyard attitude and ardent disregard for our fellow man." To the majority of pro-frackers, nature is nature on a large scale, or what we might call big "N" Nature, even when found in smaller proportions. Anti-frackers instead focus on localized pockets and unique, individually valued instances of nature found in their own communities.

In addition to the characterization of nature as vast, pro-frackers create an image of nature that is resilient. They claim that nature can bounce back from environmental degradation, reasoning that time heals the wounds of the past. In part, this view might be affected by advertising efforts by Range Resources, which emphasize the economic resilience of Pittsburgh (Rich, 2016, p. 300). Comparing the Pittsburgh of his youth to the Pittsburgh of today, Charles Clark claims:

> (12) The water is good. I can bass fish at the Point. I couldn't do that 40 years ago.

Clark attributes this recovery to nothing more than the passage of time. In a similar argument, Dan Bagley also refers to his younger years:

> (13) I was raised in Arkansas amid oil wells and gas wells in the natural state.

By referring to oil and gas wells "in the natural state," he conveys that nature encompasses industrial activity and overcomes any ill it does. Bagley continues this line of argument in his description of Pennsylvania:

> (14) I have seen that after drilling, gas wells blend with the land, and then they make money for years.

These speakers create a sense that nature will outlast industrial activity or, at least, blend in over time. These speakers disregard the invisible and lasting threats that anti-frackers worry will forever alter their communities and the natural places that they value.

When pro-frackers refer to Pennsylvania's natural lands as vast, they use synecdoche to subsume parts of the state that might not fit their wide-angled view (such as natural sites that have been degraded by industrial activity). Adrian Markocic uses synecdoche to create a broad view of nature that rejects the differential valuation of its parts. Though some Pennsylvanians may prefer to keep resource extraction far away, this requires the position that our nature *here* is more important than their nature *there*.

Industrial Places

Even more often than natural places, speakers in this hearing constructed representations of industrial places, where the most change is visible. Industrial places are referenced 141 times, ranging from the machinery of well sites to large-scale geopolitical entities characterized as predominantly industrial. Similar to natural places, industrial places are described at varying levels of specificity. Following the trend established in the previous section, pro-frackers tend to provide a more general view while anti-frackers are more specific. In presenting this wider-scale view, pro-frackers emphasize the long period of time that specific counties and states have been sites of industrial activity as well as the large scale of industrial activity in these sites, but rarely depict them in detail. Instead, they gloss over the process itself. Anti-frackers provide a closer view of the industrial process and frequently invoke the machinery of hydraulic fracturing.

Machinery: Anti-Fracker Focus

Overwhelmingly present in anti-fracker discourse are the machinery of hydraulic fracturing, from well sites to the components of these sites such as condensate tanks, frack ponds, and compressor stations. This machinery is conspicuously absent from pro-fracker discourse, which foregoes details and presents scenes broadly. Ron Slabe of Citizens Against Marcellus Pollution provides a typical example of anti-fracker discussion of well sites, in which he both discusses and evaluates the suitability of machinery for ensuring the safety of community members. Slabe says:

> (15) Marcellus deep well drilling is a destructive industrial process, which threatens the well-being of Pennsylvania residents. Its use of hydraulic fracturing pumps toxic carcinogens into the ground that, when later retrieved, are stored in *unguarded, unfenced and leaking holding ponds* that become a major source of soil and water contamination [italics added].

This evaluative description emphasizes the risk that fracturing sites pose to citizens, providing a glimpse into the restricted zones of industrial development that most audiences cannot experience on their own. Though Slabe does detail the components of hydraulic fracturing, his example does not provide an up-close encounter with an individual industrial site, which is typical of other anti-frackers, including Kevin Loughney, who was also cited in the previous section:

> (16) My brother works in the drilling industry. I seriously worry about his health and safety every day. He took me to jobsites and I saw the proximity of these wells, the containment tanks, the homes and streams and the carving of the land, the destruction of fertile farm soil. The sheer magnitude of the operation was stunning. And I really encourage you, if you get a chance, to go out and see one of these drills happening in the process—the sound, the traffic, it is overwhelming.

While Loughney details the same apparatuses of the fracking industry as Slabe, he employs synecdoche to emphasize that this drilling operation is representational of all other drilling operations. He implies that other sites will be similar. Paired with Slabe's testimony on the dangers of fracturing apparatuses, they create a broad sense of risk from individual parts. This is a typical use of synecdoche to argue that individual parts represent the greater whole.

Pro-Frackers: Expansive Activity

Pro-frackers construct industrial places on a much larger scale than anti-frackers. As I have shown above, anti-frackers focus their attention on the components of hydraulic fracturing. Pro-frackers construct an expansive view of industrial activity, often characterizing entire geopolitical areas as significantly industrial. These geopolitical areas are most often Allegheny

County and the Commonwealth of Pennsylvania, mentioned in 38 percent of pro-fracker speeches. For instance, Bernie Ulincy, a landman (a person who secures land leases for energy companies popularly depicted in the film *Promised Land*), provides evidence meant to demonstrate the expansiveness of the natural gas industry:

> (17) The number 35,000 acres has been thrown around in Allegheny County, but the reality is that there's closer to over 90,000 acres of oil and gas leases in this county.

Ulincy does not explicitly state that Allegheny County is an industrial zone. By drawing attention to an error in the estimation of the total leased land in the county, he creates the impression that Allegheny County is much more industrialized than commonly thought. He uses this claim to warn the county council that those who have stakes in the development of natural gas may seek legal compensation if this source of revenue is lost. Another pro-fracker, McHale, makes a similar claim:

> (18) Currently, the Department of Environmental Protection lists more than 1,200 oil and gas wells in Allegheny County. When I think of the increasing price of gas at the pump and the impact it has on my family's budget, I wonder just how many—or how few—barrels of crude come from local wells.

McHale and Ulincy both depict Allegheny County as significantly industrial and relate industrial activity to the financial well-being of county residents. This argument relies on the perception that Allegheny County is already heavily involved in industrial activity. McHale also values local energy more than resources that are extracted abroad. This is another common line of pro-fracker argumentation, in which local energy development is seen as a way to cripple foreign nations characterized as enemies of the United States. This will be discussed in detail in the following section.

The sense that Allegheny County and Pennsylvania are significantly industrialized is not only constructed through comments like those above, which emphasize the geographic spread of fracking; it is also created through appeals to time. By maintaining that geopolitical areas have been industrialized for many years, speakers argue for a continuation of the status quo. Mark Sommer (19) argues that the same industrial activity that is being discussed in the hearing has been going on in Pennsylvania for nearly 140 years, while McHale (20) makes a similar claim on the long history of gas extraction:

(19) Natural gas is a clean energy source produced in Pennsylvania since 1878, beginning with the Haymaker Well in Murrysville.

(20) Natural gas in the Marcellus Shale is no different than natural gas produced in Allegheny County from shallow wells for the past 100 years.

These representations of Allegheny County and Pennsylvania as industrial centers for more than a century further support the argument that the county council would be subverting the status quo if it were to ban fracking. By arguing that natural gas extraction has been present in these places for so long, they also indicate that the character of these places throughout recent history has been industrial. Advances in hydraulic fracturing technology, like the unconventional natural gas extraction methods that allow for horizontal drilling, are minimized as they stress the history of natural gas extraction.

This same strategy of creating a sense of vastness and continuity through time is used by pro-frackers to represent natural places. It relies on the synecdoche of the whole subsuming the parts. Through their use of history, pro-frackers attempt to assuage contemporary worries. Likewise, the vast and complex industrial facilities of the state subsume the arguments that fracking sites must be considered on a case-by-case basis, as some places are more fragile than others. That is, pro-frackers interpret place using synecdoche from whole-to-parts to flatten the differences that may arise from a more localized view. Anti-frackers tend toward the opposite synecdochic representation of natural and industrial places as they provide more detailed views of the local and emphasize either the unique value of nature or the specific threat that industrial machinery poses to their community and to a larger implied whole.

Home Places

Despite the fact that homes are not discussed as frequently as natural or industrial places in the Allegheny County hearing, their representation helps elucidate larger strategies of pro- and anti-frackers. Timothy Cresswell (2013) describes "home" as an "exemplary kind of place where people feel a sense of attachment and rootedness. Home, more than anywhere else, is seen as a center of meaning and a field of care" (p. 39). Some speakers tap into the connection between people and "home" that Cresswell describes to threaten council members to uproot themselves and their families to flee the perceived danger of fracking. Likewise, others represent home very broadly as beyond the physical domicile to a county, region, or state to extend their "field of care."

The construction of home as an argumentative resource manifests in three distinct ways that differentiate by scale of representation: (1) The first is a reference to a person's own home either through description or references to its relationship to industrial processes. (2) The second is a more general reference to "homes" that invokes a collective through the use of a possessive pronoun, as in "our homes" or "your homes." (3) Home is also constructed as a geopolitical space, meaning that speakers may refer to Allegheny County, Pennsylvania or other geopolitical areas as their home. The findings here align with the view of anti-frackers as focused on places that are smaller in scale and in closer proximity to the hearing than pro-frackers, who tend toward a larger scale.

Pro-frackers spend nearly twice as much time representing home in the hearing, at nearly two references per pro-fracker speech compared to one per anti-fracker. Nonetheless, because there are twice as many anti- as pro-frackers, anti-frackers make 53 percent of the total mentions of home while pro-frackers make 47 percent. In total, there are 91 mentions of home compared to 113 and 141 references to natural places and industrial places respectively.

Home as a Personal Dwelling

Though pro-frackers reference the general concept of home more, anti-frackers refer to home as a personal dwelling with greater frequency than pro-frackers at a rate of 17 references to 6. When mentioning a personal dwelling, speakers support several claims, separated by their ideological position on the issue. Anti-frackers stress the impact of natural gas on the home itself, taking a "NIMBY" stance and threatening to move away if fracking continues. For example, Jeanne McMullen claims:

> (21) The media has labeled me an environmental extremist. I am not fighting for clean air or clean water, but for the health and safety of my family and the property value of my home.

Property values are a common issue that anti-frackers champion as an ill-effect of the fracking industry. Some of these speakers, like McMullen, couch this argument solely as an issue of personal interest (the value of *my home*). This focus on personal dwellings often leads to ego-centric argumentation that focuses solely on the individual and contributes to the perception that the public sphere is in decline due to an increase in self-interested argumentation (Killingsworth & Palmer, 1992; Levasseur & Carlin, 2001).

Some anti-frackers take this individual focus even further as they threaten to leave their homes. After outlining her existing health problems, Heather Kropf states:

> (22) I don't want to live here if I'm going to be assaulted daily with continued and increased toxins. I won't survive . . . I'm probably going to sell my house and move away. I don't want to do that for this reason. I ask that you know what we're dealing with before you bring it to my backyard.

Kropf, among others, makes a direct NIMBY appeal, valuing her personal property over distant land. This type of proximity-based argument is common to anti-frackers.

Kevin Loughney, also an anti-fracker, uses his home to explain that although he understands the value of natural gas, he places his personal and spiritual needs above heating his home, creating a type of value hierarchy (Perelman & Olbrechts-Tyteca, 1969, p. 81):

> (23) I'm rational. I'm not a zealot. I realize our energy needs as a country. I heat my home and I cook my food with natural gas. But my heart, my body and my mind are fed from the land, whether it be water or soil or air.

Loughney's specific mention of his amenities transmits that he understands the value of natural gas, but, along with many other anti-frackers, he values these amenities much less than the more "spiritual" resources he locates in the land. The value hierarchy he constructs, in which he places gas at the bottom, proves compelling to the audience, as Perelman and Olbrechts-Tyteca explain in a different context: "Most values are indeed shared by a great number of audiences, and a particularly audience is characterized less by which values it accepts than by the way it grades them" (p. 81). Through his hierarchy, Loughney argues that in spite of the value of natural gas, extraction should wait until it poses no credible threat to the environment, a common line of argument among anti-frackers.

It is important to note that McMullen, Kropf and Loughney use the private space of home in a deliberative setting for a public purpose. Although, particularly for McMullen and Kropf, their personal concerns might be shared by community members, they make no explicit moves to extend them to the audience. These anti-frackers rely on synecdoche to encourage this extension of part-to-whole as their private concerns become the concern of others. At the same time, this type of rhetoric can also encourage the claims that anti-frackers are advancing the much maligned not-in-my-backyard appeal, as seen above by Markocic.

In contrast to anti-frackers, pro-frackers refer to their homes to stress that the fracking they have experienced on their own land is safe. They argue against the type of environmental threat that Loughney emphasizes, maintaining the safety of natural gas extraction as it currently exists. Eric Cesaratto, a pro-fracker, claims:

> (24) Let's try to develop the natural gas here and heat our homes. My home is heated by natural gas. I don't have a gas bill, okay, because I have a well at my house, which is safely run. And my children drink that water, and we heat our house from natural gas wells.

Cesaratto sets his personal experience with his home against claims that fracking will devalue land. He appeals to the financial value of fracking as well as its safety by stressing that his children drink the water from the wells.

By using home as a locus for their claims, Loughney and Cesaratto make parallel arguments supporting opposite perspectives. Both begin by invoking a larger spatial entity: the nation and its energy needs in Loughney's case and Pennsylvania and its energy development in Cesaratto's. Through the image of the home, they acknowledge the importance of natural gas in domestic life. While Loughney's claim shifts toward a personal, spiritual realm that remains focused on his individual needs, Cesaratto's is grounded on the practical appeal and safety of the land and can thereby be extended to a more general audience. Though both speakers focus on the personal, Loughney maintains focus on himself and his own needs while Cesaratto explicitly intends his experience to be generalized. The more constricted versus wider focus of their respective takes on the issue exemplifies a larger trend that distinguishes the rhetoric of anti- and pro-fracking groups.

Our Homes at Risk

The mention of homes in the plural is conspicuously absent from pro-fracker discourse. Only 3 instances of this more general reference are found in the hearing while anti-frackers make 12. One of the three pro-fracker references comes from the above citation by Cesaratto, which includes the imperative let's "heat our homes" with natural gas. His appeal to natural gas as heat implies that foreign sources of energy pose a threat. Similarly, when anti-frackers refer to homes in the plural, they also work to create a sense of collective threat, indicating that an entity outside of the deictic center puts what is inside at risk (Cap, 2013, pp. 74–75). That is, the naming of "our homes" identifies that community members have much to lose if fracking comes too close to where they live. This strategy is at odds with the more ego-centric approach noted earlier, which corresponds to NIMBY appeals, and is not as common in this hearing. However, it shows that participants understand and value their civic responsibility to speak to the common good.

Speakers who emphasize an encroaching threat on their homes often contrast home with industrial places. Ron Slabe gives an example of this type of appeal:

> (25) How on earth can we continue to allow condensate and compressor units to operate, drilling to occur and fracking ponds to exist a mere 200 feet or closer to our homes?

Slabe's description of the proximity of entities common to the fracking process to residential areas, which he refers to as "our homes," amplifies the sense of threat by invoking a personal connection to the audience. He implies that everyone present is at risk. A similar example comes from Dr. Judith Kaufmann, who expresses frustration that natural gas seems to lack the type of tight regulation that marks the medical profession:

> (26) Drilling companies can lure landowners with money and no disclosures as to the potential for exposure to neurotoxins, carcinogens and potential poisons of contaminated air, drinking water and soil, not to mention potential explosions and fracking fires near their homes.

Instead of broadening the sense of risk like Slabe, Kaufmann attempts to make the chemical threats of the hydraulic fracturing process visible to landowners.

As we have seen, when anti-frackers refer to their own homes or to the homes of their community members, they most often convey them as under imminent threat. Exceptions to this occur, like Loughney's mention of heating his home with natural gas. The more specific language anti-frackers use to describe home heightens the imminence of the threat they face. It also allows them to advance concrete, hierarchical values for themselves and their community, which are "no doubt, more important to the structure of an argument than the actual values" (Perelman and Olbrechts-Tyteca, 1969, p. 81).

Geopolitical Areas as Home Places

In contrast to anti-frackers, pro-frackers tend to situate home in larger geopolitical spaces like Allegheny County or Pennsylvania and to use broader language to describe it. In the hearing, pro-frackers make 64 percent of the references to home as a geopolitical space, while anti-frackers make 36 percent. Given that anti-frackers outnumber pro-frackers nearly two to one, this becomes another striking difference between the rhetoric of pro- and anti-frackers. The propensity of pro-frackers for referencing home as a geopolitical space accounts for their higher number of overall references to home over anti-frackers.

When pro-frackers refer to home in these broad terms, they are supporting one main syllogism (which is often implied in whole or in part as an enthymeme):

1: Allegheny County/Pennsylvania/"the area" is their home;
2: implied: People should protect their homes;
3: natural gas extraction is not only safe but beneficial for the economy; therefore
4: natural gas extraction is a way to protect home.

Speakers demonstrate this line of reasoning as they refer to geopolitical entities as home places.

An example of this syllogism comes from Dan Bagley of Center Rock, Incorporated, a drilling technology company. Bagley states:

> (27) I have lived in this area for almost ten years, raising my family, working within the metals and industrial sector. I have personally worked in more than 70 countries and have seldom seen the economic engine that natural gas can create if controlled well in this state, this county, this city. Prosperous communities in Texas, California, Arkansas and Colorado are growing. They have found ways to commercialize their natural gas resources in safe balance with the earth, and we are currently doing so across Pennsylvania.

Here, Bagley claims "this area" as his home, supported by his mention of family and work. He then draws from his personal experience in this industry to argue that natural gas extraction is safe, economically beneficial and worth doing across the state. Bagley could have made his argument without mentioning that he has lived in the area for ten years and that he has a family, yet he and other pro-frackers consistently draw on their time spent in "the area" to cultivate a heightened sense of concern. Undoubtedly, pro-frackers rely on home to build ethos, but why do they define home so broadly?

In some cases, the first premise of this syllogism may be called into question, as there is a perception among many Pennsylvanians that companies and their out-of-state workers are coming to Pennsylvania for Marcellus Shale. Speakers who support fracking seek to change the perception of the industry as a group of outsiders, in part, by claiming the area as home, indicating that they will work to protect it. For instance, Philip Morath focuses on the first premise of the syllogism, leaving the rest implied. Though confirming that Marcellus Shale workers are coming from out-of-state, he leaves an opening for all workers to potentially call Pennsylvania home and work to protect it. He explains:

> (28) Since I was young, I wanted to stay in the State of Pennsylvania. However, for a portion of my career, I left to work for Barnett Shale for a company in Texas. . . . I now work, live, pay taxes and buy products in Allegheny County. Many of the people I work with have a similar story to mine. They left Pennsylvania to increase their professional skills, have since moved back to use what they have learned to develop a local resource in the proper fashion. When I moved back to Pennsylvania, I may have had a Texas plate on my car and a Texas driver's license, but I've always considered myself a Pennsylvanian.

Morath leads the audience to ask themselves whether a person with Texas license plates is actually a Texan or if they could have become a Pennsylvanian. Using himself as an example, he claims Pennsylvania as his first home, explaining that he did not want to leave but was fortunate enough to return. Referring to home on the level of the state allows him to make claims not only about himself, but also about others throughout Pennsylvania who share similar stories. By naming a geopolitical area as home instead of zeroing in on his specific domicile, he speaks authoritatively about a larger geographic area with a larger population.

As we have seen, speakers use synecdoche to associate the concept of home with a sense of protection or threat. Pro-frackers stress the whole to make implications about the parts. Through syllogistic reasoning, they transmit the assumption that people will protect their homes. Defining home broadly allows them to extend this syllogism to the county and even the state levels, claiming that workers who call Pennsylvania home will therefore protect it. Meanwhile, anti-frackers define the parts to imply something about the whole. Their more narrow focus implies a more private threat. Through synecdoche, they extend this threat to other private spaces not specifically invoked in the hearing.

No matter their side, all of the speakers in this hearing who work to create a sense of place do so with deliberative ends in mind. Their use of place differs predictably depending on their ideological position. Support of fracking correlates with a broad view of place and opposition to fracking correlates with a narrow view. These differences in representation enable argumentation that either creates a sense of protection or threat.

FLAT DATA, CRITICAL REGIONALISM, AND BUILDING PUBLICS

Pro-frackers broadly paint Pennsylvania as an entity that has been home to industrial pursuits (oil, coal, steel, and now natural gas) for more than 100 years. Likewise, though contradictory, pro-frackers refer to Pennsylvania's vast natural areas with the same expansive language. Pennsylvania at once is the site of widespread, predominant industrialism and copious, resilient nature. These simultaneous views provide support for the pro-fracker argument that it is reasonable to continue down the industrial path Pennsylvania has been on for more than a century, that there is no real threat to the wilds of Pennsylvania because there is more than enough nature to withstand the effects of industrialism, and that Pennsylvania is their home, which—by virtue of being their home—they will protect.

The broad slant that pro-frackers use contributes to the "flattening" of local difference in Pennsylvania, a concept developed by Jenny Rice (2012). Citing architectural theorist Kenneth Frampton, Rice refers to "flat data" that erases "topographically singular landscapes" (p. 202). Using geologic language, Rice (2012) and Frampton (1998) describe the *tectonics* of a place, or the surface-level manifestations of large-scale processes beneath the earth's crust. Through metaphor, tectonics represent local, visible differences from place to place. "What is particularly insidious about flat data is that it smooths over the tectonics of place," Rice claims. "The light, the air, the stubbornness of any given landscape is eradicated by a construction that makes everything level" (p. 202). Here, Rice is specifically speaking about physical entities, such as a Walgreens store that is the same in Seattle as it is in Omaha. However, expanding on the work of critical regionalists (Frampton, 1998; Tzonis and Lefaivre, 1981; Powell, 2007), Rice (2012), and others (Tell, 2012; Olson, 2012; Clark, 2012; Green & Kuswa, 2012; Wood, 2012) focus on the power of rhetoric to disrupt narratives that either flatten regional difference or present it as kitsch or nostalgia (p. 203). They claim that "rhetoric . . . works against flat data" (p. 203). When applied to regions, rhetoric is a tool of critical disruption that is used deliberately to "envision and critique relationships among people and places and envision better alternatives" (Powell, 2007, p. 9). Such an enterprise is connected to movements toward social justice and inclusivity.

My findings provide a different view of the intersection of rhetoric and regionalism, derived not from an interest in positioning rhetoric as a tool of critical disruption or to ascribe to it normative value, but rather to understand the many functions of rhetoric in a site of ordinary democracy. Viewing public discourse through the lens of ordinary democracy assumes little about how rhetoric *should* be used and instead uncovers how it *is* used by citizens. Rice and her colleagues (2012) present rhetoric as a tool for critical disruption (which it most certainly can be), but pro-frackers use it to create regions, not disrupt them. Pro-frackers flatten broad areas of land to support an ideological position that industry is positive, nature is plentiful, and our home (defined broadly) will be protected. Instead of highlighting local difference, pro-frackers emphasize regional sameness. Rice (2012) and Wood (2012) recognize that all region-making is rhetorical; however, by taking a critical stance, they downplay the importance of descriptive accounts of region-making in sites like the one I have studied. Region-making is not always an enterprise of larger political forces defining the meaning of place for the less powerful. It may also be everyday citizens who seek to define regions for political power. Wood acknowledges that "our efforts to unpack the rhetoric of region may simply affirm dominant power relationships. . . ." (p. 290). Indeed, pro-frackers represent dominant power as they support the view of industry and, to a large extent, state-level politicians. Their use of synecdoche subsumes the local much more than it connects distant people into new relationships (or folds) of power (p. 290).

Anti-frackers represent place in line with the project of critical regionalists, emphasizing local difference. They describe industrial places up close, focusing on machinery and processes. Natural places are uniquely valued and, though conceived as part of a larger ecology, nature is not portrayed as the vast, unbroken entity it is by pro-frackers. Home places are mostly individual homes or "our homes" within Allegheny County, while pro-frackers call home geopolitical entities like Allegheny County or Pennsylvania. The anti-fracker focus on the uniqueness of place in the small-scale disrupts the region-building efforts by pro-frackers, reminding the county council that their interest as a political body is in Allegheny County and not Pennsylvania as a whole. Rice (2012) states that the critical regionalist effort is one of "making people feel at home and reflecting the current conditions of culture in that area" (p. 212). Anti-frackers participate in this effort by identifying perhaps overlooked places and ascribing value to them. By emphasizing their uniqueness and meaning, they indicate that the loss of these places is an unacceptable risk. Instead, they should be protected and shared. Their use of synecdoche builds regions through implied rather than explicit language. Although others may recognize themselves and their concerns in the speeches of anti-frackers, they do not explicitly attempt to connect their experience with others. As a result, anti-fracker rhetoric sometimes comes across as a mere airing of personal grievances.

At the same time, the individual tone of anti-fracker representations of place may run counter to the project of public building. In particular, their representation of homes advances an ego-centric view of public responsibility through which they prompt their representatives to pay attention to and act on their individual concerns. As we have seen, some even threaten to move away if their concerns are not answered. This raises questions about whether the rhetoric of critical regionalism, which seeks to disrupt narratives that flatten individual concerns or the concerns of communities, may in some cases promote an individualism that is counter to the common good. Gerard Hauser (1999) explains that when common affiliation is lost from a society (for Hauser based in memories of the past), "it becomes disoriented, it begins to lose its capacity for common action, and its members find the meaning of their individual lives impoverished" (p. 119). The affiliation between members of the public present in the Allegheny County hearing is not so much built on memory as it is on the shared world that they live in and the representations of place that evoke common concerns. Hauser states, "A public's emergence is not dependent on consensus but on the sharing of a *common* world, even when understood and lived differently by different segments of society" (p. 69). Critical to this account is that the world people share becomes common by way of discourse that invokes places of recognition and the values found therein.

For both pro- and anti-frackers, the trouble they may have in creating the common world necessary for public building is an issue of scale. Anti-frackers may focus too narrowly to allow for identification among community members, sometimes excluding others entirely by focusing solely on their own concerns and never explicitly reaching out to their fellow citizens. Pro-frackers often speak about state-wide or nation-wide issues that may resonate with their fellow hearing participants, but may also miss the mark for the particular purpose that brings them together: to determine the course of action within Allegheny County. Though publics are conceived as reticulate by Hauser (1999, p. 12), the particular public that is called into being for the purpose of determining whether or not to frack within Allegheny County may suffer from a lack of focus at the county level itself. This may be in part due to the ill-defined nature of the hearing. Chair Martoni does specify in the beginning of the hearing that comments are to be restricted to Allegheny County, but this is not mentioned again even as citizens frequently discuss issues that go well beyond the jurisdiction of the County Council or that dip into their individual problems. Despite the ways these participants miss the mark, speakers make evident the importance of place to them and to their communities. As they stretch out beyond the county-level scope of the hearing to regional and national perspectives or narrow their focus within their own personal sphere, they attempt to attach meaning to the places they do share by relating the part to the whole and the whole to its many parts.

CONCLUSION

In this chapter, I offer a methodological approach to place-based arguments. This approach, suited primarily to environmental controversy, divides general and specific place references into six categories. The most useful of these categories for this study have been nature, industry, and home, each conceived as proximate or distant to the deictic center. Speakers also positively or negatively evaluate the land, the merits of fracking, and each other through these depictions of place. Using this approach, I find that speakers regularly use synecdoche to represent their perspective on the issue. These insights help us understand how rhetoric builds our affiliations to place. Such place-based affiliations define commitments, values, and bonds between ourselves, our communities, and the land. In the next chapter, I explore how rhetors share experiences in place with others so that they may apprehend scenes they have not witnessed themselves.

NOTES

1. The subject of discourse in this chapter is in all cases a place. Place takes the role of setting and subject as people actively construct the meaning of places for their audience.
2. I also considered, using Piotr Cap's (2013) Spatial/Axiological/Temporal framework, the discursive construction of impending threat. Cap's coding scheme applied to this hearing provided an interesting insight, but seemed to be getting ahead of what I sought to learn from this hearing. Surely, many people refer to places to create a sense of threat coming from outside to inside of the deictic center, but this excluded many argumentative uses of place. At this point, I returned to my original catalog of all instances where places are mentioned in the hearing.

Chapter 3

Constructing the Vicarious Experience of Proximity in a Marcellus Shale Public Hearing

In this chapter, I move from a broad study of place-based arguments in an Allegheny County public hearing on hydraulic fracturing to an analysis of the specific topos of the rhetoric of place. As mentioned earlier, Carbaugh and Cerulli (2013) argue that communication is "doubly-placed"; it is "located in place, and, it shapes our senses of our places" (p. 7). They highlight one type of discourse that achieves this effect: the verbal depiction of land (p. 14). When writers and speakers depict the land, not only do they familiarize audiences with parts of the physical world they may not know, they also draw attention to "larger systems of practice" and "symbolic contexts for examining this place" (p. 15). Through the depiction of scenes, audiences enter a more proximate relationship with place. In turn, place-based connections form among people and the processes by which they form come into relief.

Place-based connections are particularly important in hydraulic fracturing discourse. As shown in the previous chapter, through place-based appeals, citizens offer contrasting worldviews. They represent places as unique parts of a valuable whole or as a resilient whole made of many representative parts. These differing depictions carry meaning and can impact how people experience shared places in their communities. Depictions of place are also a useful tool to establish ethos on accounts of industry, nature, and home. Citizens use place-based appeals to show they have knowledge alongside those with industrial or academic credentials (who also rely on place-based rhetoric). In other words, it helps them justify their inclusion in public hearings and support their perspective. Place-based appeals allow them to shape perceptions of the important locations in their lives and create a sense of proximity that brings their audience in closer contact with these places.

I begin with an example of place-based discourse that creates a sense of proximity. Part of this excerpt you will recognize from chapter two as an example of synecdoche that characterizes specific industrial activities as representative of threats across Pennsylvania. The speaker, Kevin Loughney, provides a first-hand account of a well site being drilled.

> The many camping trips I take to state parks and forests all over the state had us bear witness to the broad and negative effects of drilling. My brother works in the drilling industry. I seriously worry about his health and safety every day. He took me to jobsites and I saw the proximity of these wells, the containment tanks, the homes and streams and the carving of the land, the destruction of fertile farm soil. The sheer magnitude of the operation was stunning.
>
> And I really encourage you, if you get a chance to go out and see one of these drills happening in the process—the sound, the traffic, it is overwhelming. The frack that I saw was aided by tanker after tanker of liquid nitrogen. They said it was a small well and that it was cleaner than using more fracking fluid. Liquid nitrogen into our ground, directly into our ground. I was awed, sickened and sad to see these things going on in highly populated areas, state lands and our forests. Please hold these companies to task, for me and the average Pittsburgher and county resident depend on it. Don't take their money—without first being sure this is what you want in our backyards.

Loughney's account provides audiences with a sense of proximity to the scene he has witnessed by presenting and evaluating a first-hand description of a place. Speeches like Loughney's spark two questions:

1. *How is a sense of proximity created in discourse?*
2. *What does the vicarious experience of proximity achieve in a site of ordinary democracy?*
3. *How can we put this knowledge to use in future advocacy efforts?*

To answer these questions, I look at all of the contributions to the hearing in which a specific geographic location is depicted for the purpose of bearing witness to the experience. The rhetorical effect of these appeals is to provide the audience with a vicarious experience of a place. In doing so, they control the representation and lead their audience toward a certain evaluation. The speaker assumes that if the audience were to experience that place first-hand—to be proximate to it—their positions on the issue would match. Due to the vicarious nature of this appeal and the creation of a sense of proximity, I call this "vicarious proximity" (Mando, 2016).

This concept and the coding scheme I present below has been independently applied by Scarff (2021) in an investigation of Mountain Valley Pipeline citizen comments. Though she broadens the criteria for vicarious proximity, she finds that "vicarious proximity appeals overwhelmingly mapped onto the corpus of citizen comments" (p. 4). Scarff's analysis supports the explanatory power of this concept when applied to citizen's environmental discourse.

Vicarious proximity appeals have four defining dimensions or *critical features*. These features are *critical* in the sense that they are essential components of vicarious proximity appeals. This is not to say that all speakers in the hearing make use of these features, but the speakers who verbally create an experience of a place for their audience do use them. The four critical features are the following: (1) *focus on specific place through place-names and space-relational descriptors*, (2) *visual vividness*, (3) *first-person reporting*, and (4) *bearing witness*. The first three of these features are discursive dimensions of vicarious proximity appeals while the fourth, bearing witness, is an evaluative stance that is taken in reaction to how the place in focus has been transformed. Loughney, above, demonstrates this by lamenting the "destruction of fertile farm soil." These four critical features can serve as identifying characteristics of this appeal for analytical purposes as well as guidelines for the production of such appeals.

By using the discursive dimensions of vicarious proximity appeals as a lens to compare the discourse of pro- and anti-frackers, I will support two major claims about the ways these groups use place as an argumentative resource. Anti-frackers tend to create one focused scene for their audience with high levels of detail. Pro-frackers, instead, tend to jump from scene to scene and are less detailed. This results in two ways of experiencing place conveyed through an analogy: pro-frackers give their audience an aerial view with less detail but that covers more area while anti-frackers take their audience on foot to provide an on-the-ground view of one detailed scene. This finding has implications on audience involvement and the role of place-based rhetoric in democratic participation.

"PRESENCE IN THE WORLD OF THE TEXT"

Vicarious proximity appeals fit George Lakoff's (2010) call for environmental communicators to tell stories, look beyond quantitative data, and make use of visuals by presenting first-hand experience through place-based narratives (p. 79–80). Readers may assume that arguing through detailed depictions of place, often natural settings, follows a long tradition in environmental writing by those such as Henry David Thoreau, John Muir, Aldo Leopold, and Rachel Carson, but this is not the full picture. Vicarious proximity appeals are used nearly equally by those in support of and in opposition to hydraulic fracturing. Because they are used broadly and are *not* solely the purview of environmentalists, I present vicarious proximity as a *topos* of place-based rhetoric. Due to their ubiquitous nature, understanding how they are used by pro- and anti-frackers builds on the insights of the previous chapter to provide a more focused view of the constraints on speakers in public hearings about environmental issues as well as the different ways place is used as an argumentative resource.

I find that public hearing speakers respond to the common rhetorical problem of affecting a political situation by verbalizing experiential knowledge of a place. This experiential knowledge is specifically gained through time spent in a place that they have witnessed transform, whether it is the destruction of what had been a pristine forest or the new-found prosperity of a formerly defunct rural town. They present this transformed place as a scene for the audience to visualize and evaluate. The rhetorical purpose of these appeals is to communicate the consequences of a type of land-based intervention and to invite audience members to actively share in this experience by visualizing the scene. Through visualization, they achieve vicarious proximity to the transformed place, a process that is similar to what rhetoricians know as *ekphrasis* (the verbal depiction of visual phenomena) or Perelman and Olbrecht-Tyteca's (1969) *presence*. What sets vicarious proximity apart is the use of place as the primary carrier of meaning. This conception of place departs from Labov's (1972) narrative model, which privileges temporality, and follows Schiffrin (2009), who emphasizes the general role of place in narrative and focuses on narratives that create place identity.

By creating an experience of a place through text, vicarious proximity appeals are a type of scenic writing. Kaufer and Butler (2000) claim that "done well, scenic writing causes the reader to feel present in the world of the text" (p. 85) and that "the scenic writer coaxes the reader to feel as if the words are functioning as her own eyes and ears" (p. 86). This feeling of "presence in the world of the text" is the vicarious experience of proximity to a particular place that provides the persuasive force of these appeals. Tannen (2007: p. 27) refers to this same force of scene-making as an *involvement strategy*. Especially notable is that these longer-form genres are found in three-minute public hearing speeches.

Ernest Bormann's (1972) scholarship on rhetorical vision, using his fantasy-chain analysis, illuminates a most important aspect of place depictions, namely that group imagining can make "commonalities public" (p. 397). This work is grounded in an analysis of group communication in which individuals seek a common purpose. Bormann, citing Robert Bales, argues that common ground is created among individuals through the sharing of visions. Each member participates in the creation of a singular vision, which can then reach increasing levels of publicity and importance.

> Once we participate in the rhetorical vision of a community or movement, even if we keep an aesthetic distance, we have come vicariously to experience a

way of life that would otherwise be less accessible to us, we have enlarged our awareness, we have become more fully human. (Bormann, 1972, p. 407)

I believe that vicarious proximity appeals in a public hearing function similarly, with the difference being that speakers advance competing visions of places of importance to the community. The vision these participants create is not a singular vision as in Bormann's research, but rather a sustained negotiation of place. The result of an enlarged awareness remains a good of this process.

The content of both rhetorical vision and vicarious proximity appeals is described by Bormann as "characters, real or fictitious, playing out a dramatic situation in a setting removed in time and space from the here-and-now transactions of the group" (p. 397). Bormann illuminates the rhetorical effect that I aim to communicate about vicarious proximity appeals, as audience members are (at least momentarily) transported from a chamber in the Allegheny County Council building to a devastated woodland, a bustling town rising from its former decrepitude, or a well site either leaking fracking fluids from containment ponds or functioning in a clean and orderly manner. When speakers advance these appeals, they verbally depict visual phenomenon they have experienced for others, inviting them to share their vision not only of the place described but of its future. Similar effects produced by visual imagery such as photographs, paintings and video have been studied in environmental communication (Dobrin & Morey, 2009; Peeples, 2011); however, the verbal depiction of visual imagery has received less attention.

When a group shares a rhetorical vision, they do not control the way that vision is perceived. The audience's interpretation of the image varies. They do not apprehend it in the same terms. Still, scholarship on ekphrasis stresses the apprehension of the object described. Rhetorician Susan Jarratt (1998) defines ekphrasis as "a verbal depiction of visual scenes" (p. 48). Rhetoricians who use this term (Jarratt, 1998; Perry, 2013) leave the latter half of the definition open for wider application. Literary theorists such as James Heffernan (1993) and Mack Smith (1995) restrict ekphrasis to the verbal representation of visual art. An example of this is poet William Carlos Williams's *Pictures from Breughel*, which Smith describes as "a work of language remaking visual art" (p. 154). Despite the more constricted literary history of the term, rhetoricians such as Samuel Perry (2013), who studies the use of ekphrasis in Billie Holiday's "Strange Fruit," broaden ekphrasis to describe visual phenomena depicted in language. Perry writes that Holiday incites her audience to experience lynching by "leveraging" the scene "in ways that appeal to multiple senses" (p. 452). This verbal recreation of a scene puts audience members into a new relationship to an act that may otherwise be unknown, facilitating critical engagement.

Vicarious proximity appeals function similarly in that they are the verbal representation of a visual experience, but unlike ekphrasis, accuracy of apprehension is not paramount. As long as audience members visualize the scene described and are put into a critical relationship with it, the work of the vicarious proximity appeal has been done. What is most important about rhetorical vision and ekphrasis is that the experience created for the audience is one that audience members may circulate themselves. This is evidenced by vicarious proximity appeals that take shape as both first- and second-hand accounts. The power of this form of representation is in a rhetor's ability to present a scene that becomes real enough for those who apprehend it so that they may recount it to others. This circulation, in turn, contributes to the formation of a public.

Vicarious proximity depends upon the intersubjective relationship between speaker and audience. Though in the setting of a public hearing there is very little interchange between these two parties, by way of vicarious proximity appeals, speakers ask their audience to occupy a space of intersubjectivity. This allows for the co-creation of place. Speakers describe and even explicitly invite their audience to imagine along with them places that have shared meaning in their community. This invitation entails what Perelman and Olbrechts-Tyteca (1969) refer to as "communion" (p. 51, p. 178), but may also be described as "rhetorical distance" (Hunsaker & Smith, 1973; Crowley & Hawhee, 1999; and Osborn, 2004). While Tannen's (2007) perspective on scene-making emphasizes *how* involvement is created, the rhetoricians cited above are more focused on answering *to what end*? For rhetoricians, the space between speaker and audience relies on the assumption of a shared set of values that can be activated for a persuasive effect. The power of effectively negotiating rhetorical distance, according to Osborn (2004), is to enable the manipulation of "symbolic space" through which "the rhetor can augment or diminish appropriate *feelings*, enhance *identification*, bestir *action*, and *celebrate* values and the icons that personify them" (p. 33, emphasis in original). This symbolic space is described by Crowley and Hawhee (1999) as a metaphor that represents the "degree of physical and social distance" between speakers and audiences (p. 182).

Osborn defines rhetorical distance in Dr. Martin Luther King's "Letter from Birmingham Jail" as a distance between audiences; one is clearly addressed and the other is not, though remains the actual audience of the text. In Osborn's view, rhetorical distance entails the creation of moral absolutes that create space between King's movement and his opposition. Vicarious proximity overlaps with accounts of rhetorical distance in some regards and differs in others. The rhetorical effects of enhanced feeling, identification and action, as described by Osborn, are similarly targeted; however, with vicarious proximity appeals these effects are not accomplished by increasing or reducing metaphorical distance between rhetor and audience. Vicarious proximity appeals require a third entity: physical place. The aim of vicarious proximity appeals is to reduce the distance between audience and place, while the rhetor establishes him or herself in closest proximity to that place. The effect can be imagined as the rhetor inhabiting a place and beckoning the audience to come nearer and nearer to gain a closer view. Though this appeal is markedly different from the rhetorical distance the aforementioned authors describe, the ultimate effect of rhetorical distance and vicarious proximity are "strategically selected accounts of situations that will induce listeners to respond and act" (Osborn, 2004, p. 31). In the case I have studied, the action requested of audiences is not to change their perception of the rhetor; that is only tangential. What is desired is a more proximate relationship between audience and place that leads to action.

Proximity plays an important role in our understanding of issues that have significance for a community, whether it is geographic or symbolic. Scottish Enlightenment philosopher of rhetoric George Campbell claims that people are more interested in events that have happened recently in contrast to those in the distant past, but even more important is their "local connexion" (Golden & Corbett, 1990, pp. 218–219). He explains:

> With how much indifference, at least how slight and transient emotion, do we read in newspapers the accounts of the most deplorable accidents in countries distant and unknown! How much, on the contrary, are we alarmed or agitated on being informed that any such accident hath happened in our neighborhood, and that even though we be totally unacquainted with the persons concerned! (Golden & Corbett, 1990, pp. 218–219)

Geographic proximity to an issue has been shown to not only increase interest in issues, but also authoritativeness, suspicion of outside voices, and tendencies toward moderate views (Blommaert et al., 2003). In addition to garnering interest, proximity is also created in discourse to increase a sense of threat (Cap, 2013).

Proximity also plays a large role in public argumentation and media representation. Consider a particular form of proximity appeal that has become a common term to describe the situation of local people when they reject the placement of what they see as a negative presence in their community. As seen in the previous chapter, the *NIMBY* appeal can be a powerful move to relate one's home to a public and place-based issue. This has been notably studied in Katz and Miller's (1996) analysis of the siting of a low-level nuclear power plant. Studies like these show the relevance of place and proximity to issues of power and domination.

In media representation, van Dijk (1985) as well as Bednarek and Caple (2013) have provided detailed insight into what makes a story newsworthy.

Proximity is considered a "news value" that affects the interest of a story to an audience. Likewise, proximity has also been shown to affect not only *what* is reported but *how*, as distance influences the reporting of an event (Johnstone & Mando, 2015). "Perceived proximity" constructs journalistic authority (Zelizer, 1990), contributes to a sense of national consciousness (Volkmer, 2008) and facilitates organizational communication (Wilson et al., 2008). Volkmer finds that the globalization of media creates new "event spheres" that blur the boundaries between nation-states and create a "globalized public space" (p. 92).

In particular, Zelizer (1990) discusses the role of proximity in establishing journalistic authority. She employs an analytic framework that considers the way the following conditions influence reporting: (1) place of event, (2) place of journalist, and (3) place of broadcast. She finds that "this 'real' spatial connection often cannot be established" and "presentational practices" help establish a sense of proximity where often there is none (p. 40). This is accomplished by relying on audience assumptions that, for instance, the reported place of the event is also the location of the reporter. By focusing on the event's location, reporters can withhold their own location until the very end of a report and maintain the audience's assumption that they are near the event and therefore more authorized to report on it. Zelizer's finding that the construction of proximity establishes authority in the news media is similar to the way that participants in the Allegheny County public hearing utilize their proximity to a fracturing site for strategic purposes.

Proximity has also been discussed in phenomenological terms by David Lynch (1998). In Lynch's *Rhetoric Society Quarterly* article, "Rhetorics of Proximity: Empathy in Temple Grandin and Cornel West," it is linked to empathy and the process of understanding across difference. For Lynch, the rhetorics of proximity "seduce us into ambiguous social spaces" so that proximity does not relate to physical space or geography, but rather represents the metaphorical distance between two people as they attempt to understand one another when full comprehension is not possible (pp. 11–14). Though Lynch helps us understand the complicated process of empathy, I diverge from his work and take a new approach to the rhetorics of proximity that foregrounds the construction of spatial proximity in discourse to establish common ground and facilitate arguments for the common good. By doing so, I attempt to provide a discussion of both *how* such rhetorical visions garner the involvement of their audience and *to what end* they are used in a deliberative arena.

VICARIOUS PROXIMITY AS UNIT OF ANALYSIS

The data used in this analysis comes from the Sargent's court reporter's transcript of the March 10, 2011, Allegheny County Public Hearing on

Marcellus Shale. In contrast to the last chapter, I now narrow my focus onto all instances in which speakers verbally depict places, providing a descriptive account of a place. This data takes the form of short adjective with noun phrases (e.g., devastated woodland) and longer narrative appeals that depict detailed scenes. I approach my data with the general principles of grounded theory (Glaser & Strauss, 2009), developing questions to elicit meta-data and a coding scheme through multiple close readings of the corpus that allow insights to emerge from the texts. I treat the single hearing as a corpus of 85 separate speeches. Through the place-based rhetorical analysis coding scheme I used in the previous chapter, I familiarized myself with all place references in this text. That analysis revealed the particular discourse move by which speakers situated themselves in narratives through which they depicted places. Identifying all instances of this discourse resulted in 36 speeches that included the 14 I categorize as vicarious proximity appeals. I generated a list of 19 questions common to rhetorical and discourse analysis to elicit categories within the texts. These questions ranged from the professional association of the speaker and their stance on the issue to how metaphors, directional language and naming are used by the speakers. I also answered questions about how phenomena were represented, such as nature, human action, power, technology, time, and topography. Noting commonalities among speeches, these broad questions led to the emergence of the specific representation of place I refer to as vicarious proximity. I identified this discourse in 14 speeches from the hearing that included the 4 features defining these appeals and separated them for another round of analysis. I then differentiated the use of these appeals by those who support versus those who oppose hydraulic fracturing. This fine-grained view of each critical feature is explained in the corresponding sections below:

How do people identify themselves?

a. What is the business association or organizational association of the speaker?
b. Where do speakers say they are from?
c. Is their account first-hand or otherwise?
d. Do they side with pro- or anti-frackers?
e. How do speakers construct ethos?

How do speakers represent places?

f. What specific places are mentioned by the use of proper nouns?
g. What general places are mentioned?
h. How are places depicted?

i. How precisely are geographically specific places represented?
j. How is directional language used? How do people represent the distance between the hearing and their homes? Between their homes and well sites? Between Pittsburgh and other places?
k. How is nature represented? Do people take an eco-centric, homo-centric or ego-centric view of nature?
l. How is human action represented within places?

This heuristic approach allows the analyst to move from qualitative questions to coding categories that have emerged from the data. The process moves from reading texts in full to generating and answering questions to developing codes that allow for quantitative insights. Through multiple readings, revisions to my coding scheme, application of the codes, and counting the results, I arrive at four features that are common to a subset of these place-based narratives. These four features are the defining rhetorical elements of vicarious proximity appeals: (1) *focus on specific place through place-names and space-relational descriptors*, (2) *visual vividness*, (3) *first-person reporting*, and (4) *bearing witness.*

With these four critical features, I developed another round of questions specific to each feature to allow me to differentiate the ways pro- and anti-frackers employ each. For (1) *focus on specific place through place-names and space-relational descriptors*, I am interested in the types of specific places that were mentioned as represented. In (2) *visual vividness*, I asked: What is depicted? What is the strength of the depiction? What is the length of the depiction? How is it depicted? I also draw from Bierwisch's (1996) concept of intrinsically spatial language. In the (3) *first-person reporting* section, I ask: implicit or explicit? What references to time are made? What references to place are made? For whom do they speak? In the last section of analysis, (4) *bearing witness*, I ask whether this was implicit or explicit and rely on the results from my previous sections of analysis to support the claims I make. I also make note of explicit evaluation versus strategies of implied evaluation.

This process has allowed me to investigate the use of place in a site of ordinary democracy and, specifically, to better understand the role of descriptive accounts of place gained by first-hand experience as an experience of vicarious proximity. This fine-grained view of each critical feature is explained in the corresponding sections below.

Establishing Vicarious Proximity

The four critical features of vicarious proximity appeals are the following: (1) *focus on a specific place through place-names and space-relational descriptors*, (2) *visual vividness*, (3) *first-person reporting*, and (4) *bearing witness.*

In addition to these, vicarious proximity appeals also regularly include four features that are not critical to the construction of the appeal, but commonly co-occur. I refer to the following as variables: (a) *legacy or heritage*, (b) *shared place/common place*, (c) *analogous situation*, and (d) *invitation to visualize*. Vicarious proximity appeals include all of the four critical features and may include any combination of the variables, which provide depth to appeals but are not necessary features. By comparing the detail given to each of the essential features and the combination of variables included, I demonstrate differences between the discourses of anti- and pro-frackers.

To get a sense of what each of these critical features and variables mean and how they function, I will first turn to a representative example, John Muir's "The Hetch Hetchy Valley."[1] Though my primary focus is on public hearing speeches, Muir finds himself in a similar rhetorical situation to many of those at a Marcellus Shale public hearing. The relationship between Muir's piece and the speeches we encounter manifests how the constraints of a public hearing shape these appeals as well as the prevalence of vicarious proximity appeals outside of public hearing rhetoric. After showing how Muir constructs a vicarious proximity appeal, I will detail the range of possible moves that rhetors make in such appeals via the Allegheny County public hearing.

John Muir and "The Hetch Hetchy Valley"

In "The Hetch Hetchy Valley," John Muir (1908) provides examples of vicarious proximity appeals that will serve as a guide for our exploration of public hearing rhetoric. In this piece, Muir seeks to recreate sites that would be persuasive enough to gain support for the protection of Hetch Hetchy if his audience were able to see them first-hand. In this section, I will show how Muir creates vicarious proximity appeals by not only bearing witness but also via the remaining critical features of vicarious proximity appeals (1–4) and their variables (a–d).

In contrast to the perspective we will take, Christine Oravec (1981) provides insight into Muir's nature writing prior to "Hetch Hetchy." In these pieces, Muir evokes a *sublime response* "to recreate in his audience the sensation of mountain grandeur" (Oravec, 1981, p. 248). Oravec claims that Muir achieves a sublime response by providing detailed descriptions of dramatic natural scenes in which he foregrounds the spectacular and backgrounds human involvement, which creates a sense of distance and smallness before a grand scene (p. 249). In "Hetch Hetchy," I argue that Muir does not create a sublime response to make his audience feel small and his subject feel important; rather, he rhetorically locates his audience within the scene so they may bear witness along with him. The difference between Muir's earlier work and "The Hetch Hetchy Valley" is that Muir is responding to the threat of an impending change within the valley instead of presenting the grandeur of a timeless scene, as he did in his arguments to establish the Yosemite Valley as a national park.

In the essay, Muir explains that he opposes the creation of a dam in Hetch Hetchy Valley. He vividly depicts the valley as he has experienced it, focusing on creating a representation of the place itself before it was dammed and flooded. He bears witness to the setting in two ways. First, he recalls the grandeur of the valley as it is; second, he hypothetically bears witness to the valley as if it were dammed. By including these critical features and others that I will detail, much of Muir's essay can be considered a vicarious proximity appeal. First, I will detail the critical features:

(1) *Focus on a Specific Place through Place-Names and Space-Relational Descriptors*

The first critical feature of any vicarious proximity appeal is a focus on a specific place. Beyond a multitude of references to the Hetch Hetchy Valley, Muir also takes care to name specific features within the Valley. For instance, Muir poetically describes one of the largest waterfalls when he says "Looking northward across the Valley you see a plain, gray granite cliff rising abruptly out of the gardens and groves to a height of 1800 feet, and in front of it Tueeulala's silvery scarf burning with irised sun-fire." Muir invokes the Tueeulala waterfall without explicitly mentioning that it is indeed a waterfall. Using its name, he assumes that readers are familiar with it. Through his description, even those who did not know Tueeulala come to understand the waterfall as shared knowledge. This tactic can be considered an invocation of the vicarious proximity variable (b) *common place*, as Muir's assumption of familiarity constitutes readers as insiders.

Beyond the place name, Muir also geospatially locates Tueeulala by providing cardinal directions with reference points to situate the waterfall within the larger context of the Valley. In vicarious proximity appeals, concrete place is invoked more often through place names than it is through geospatial location. Muir, however, includes both, as he has the luxury of presenting his appeals in the form of an essay rather than a two-minute speech.

(2) *Visual Vividness*

As evidenced in the previous example, Muir depicts the scenes he invites his audience to visualize with great detail. He uses visual metaphors such as "Tueeulala's silvery scarf burning with irised sun-fir" to capture his audience's imagination, but also includes more pedestrian descriptions such as that of the "gray granite cliff rising abruptly out of the gardens and groves to a height of

1800 feet." These descriptions serve to bring an audience in contact with the natural phenomena he wishes to preserve. However, Muir goes further than this and includes vicarious proximity variable (d) *invitation to visualize*. This variable is an explicit invitation for the audience to view themselves within the scene Muir creates. He accomplishes this by saying, "Imagine yourself in Hetch Hetchy on a sunny day in June, standing waist-deep in grass and flowers (as I have often stood), while the great pines sway dreamily with scarcely perceptible motion." This explicit invitation prompts the audience to engage with the physical descriptors Muir provides and to co-create the experience by imagining other sensory details such as the warmth of the June sun on their skin.

(3) *First-Person Reporting*

Vicarious proximity appeals depend on first-hand accounts of an experience known through multiple modes (verbally, visually, emotionally, and/or conceptually) that are processed as abstract schema (Flower & Hayes, 1984, p. 120). This knowledge is gained through experience in a particular place that is taken to be representative of a change. Because of this, first-hand experience is often put into a relationship with time. Rather than reporting on one isolated event, those who share their experiences in vicarious proximity appeals often situate themselves as having knowledge of a place gathered over a long period of time.

Muir can be seen referring to his first-hand experience in this way in the parenthetical remark from the aforementioned passage, "as I have often stood." Muir is claiming not only that he has seen this himself, but that he has *often* been in this place, observing the scene he describes. This adds force to his observation as he positions himself as an authority on that place and the changes it has undergone.[2]

(4) *Bearing Witness*

While the critical feature of first-hand experience emphasizes an experience in a particular place, *bearing witness* is focused on the experienced change. In this, rhetors may draw attention to a change, almost always in relation to a physical place, which they view as positive or negative. They then communicate the consequences of a land-based intervention. They interact with their audience in "a rhetorical space of intersubjectivity," as Wendy Hesford (2004, p. 105) describes. By viewing *bearing witness* as such, we can understand this feature to be an interface through which rhetors and audiences may confront each other over a public issue. It is central to vicarious proximity appeals because it is what brings them into public deliberation. The evaluation in *bearing witness* forms alliances, challenges perceptions, and provides a space for contestation.

In vicarious proximity appeals, rhetors most often bear witness to completed or ongoing acts that they refer to either in the past or present tense. However, it is possible that rhetors may also bear witness to a hypothetical future that they purport will come to be if a certain path is taken, as John Muir does in his description of Hetch Hetchy Valley as if it had been dammed: "Thus the Hetch Hetchy dam-lake would only be a rough imitation of a natural lake for a few months of the spring months, an open sepulcher for the others." Muir points out that damming the Valley would leave it void of water for much of the year. The picturesque valley he has previously created for our visualization now becomes an "open sepulcher."

These are the four critical features of any vicarious proximity appeal. Now, we will turn to the public hearing to investigate the components of these features in more detail and in their particular context as public hearing rhetoric. The variables (a–d), two of which I have mentioned above, will also be defined.

Critical Feature #1: Focus on Specific Place through Place-Names and Space-Relational Descriptors

As we have seen in Muir's "The Hetch Hetchy Valley," by focusing their vicarious proximity appeal on a specific place, speakers provide context. Less obviously, the way speakers use descriptive language determines whether and how much the audience is asked to co-create the scene. Focusing on a specific place can be accomplished in two ways: (1) speakers may reference a concrete place name and, (2) they may locate the *origo* (i.e., the reference point of a deictic relationship) of their vicarious proximity appeal by use of descriptors that signal spatial relations (e.g., "just beyond the airport") and ordinal directions (e.g., "in the northwest corner of the county").

Specific places mentioned in the hearing take the form of political units (e.g., cities, counties) such as "Clearfield County," geographic features such as the "Monongahela River," or enterprises such as the "Greenfield Coke Plant." The majority of participants in the public hearing refer to political units over geographic features. For instance, Mel Packer reads aloud an email from an unnamed wildlife biologist who reports on a visit to a nearby forest with unspecified others, "We were just up in Moshannon State Forest in Clearfield County." Packer's email provides both a geographic feature and a political unit to situate his audience into a geospatial context.

Others, such as Kate St. John, provide appeals of a situation analogous to the subject of the hearing, situated in close proximity to the hearing itself. St.

John provides the following narrative: "Let me tell you a similar story from my neighborhood. I live in Greenfield, one mile from the site of the former Hazelwood Coke Plant." St. John specifically mentions the Pittsburgh neighborhood, Greenfield, and situates it in relationship to the Hazelwood Coke Plant. In doing so, she provides context for her vicarious proximity appeal, which is intended to help her audience visualize the specific location of her narrative. St. John also makes use of the variable (c) *analogous situation* to draw a comparison between her experience of a pollution-causing enterprise and the effects it has on those nearby. The proximity of the situation she recounts to the hearing itself also helps St. John argue for the applicability of her comparison.

Though Packer and St. John are both against hydraulic fracturing in Allegheny County, the focus on specific places and use of proper nouns and spatial descriptors occur in near equal numbers by those on both sides of the issue. Out of the six anti-fracking speakers, there are 20 specific mentions of place as opposed to 22 from the eight on the supporting side. Pro-fracking speakers locate their appeals at a state (36%) or county (32%) level more frequently than anti-frackers. Anti-fracking speakers situate their appeals in states 15 percent of the time and counties only 10 percent of the time, instead preferring scenes at a smaller scale.

Along with the trend for pro-fracking speakers to be less descriptive (discussed below), this suggests that pro-frackers spare specifics to paint a broader scene. For instance, Ed Valentas describes the prosperity he has seen as the result of revenue from the gas industry: "Now the economic impact of oil lease, bonus monies in Washington and Butler Counties, you go down there, there's . . . ." Valentas situates his appeal at the county level, implying that the experiences of both counties are analogous and supporting his claim that Allegheny County would behave similarly. The speaker also situates the *origo* of his appeal in relation to the hearing in Pittsburgh (north of Washington and Butler counties) when he says, "You go down there . . . ." When Valentas uses deixis (there) without a specific referent, his comment applies broadly. Audience members may bring their own knowledge of Washington and Butler counties to co-create the scene of prosperity that Valentas presents. His lack of geographic specificity is a trait shared by other pro-frackers, who provide their audiences with a more aerial view. While their claims about fracking can cover broad areas, anti-frackers create scenes that are smaller in scale and more detailed.

Specific place is the first of the four critical features found in vicarious proximity appeals and has been shown to appear in near equal levels among those who support and oppose hydraulic fracturing. This critical feature sets the context for the vicarious experience of the audience member and influences the degree to which the audience co-creates the visualization.

Table 3.1 Counts of Specific Place References across Vicarious Proximity Texts

Specific Place	*Pro-Fracking (8 texts)*	*Anti-Fracking (6 texts)*	*TOTAL (14 texts)*
Country	0	1	1
State	8	3	11
County	7	2	9
City	2	2	4
Municipality	3	2	5
City Neighborhood	0	6	6
Natural Place	2	2	4
Enterprise	0	2	2
TOTAL	22	20	42

Critical Feature #2: Visual Vividness

Tannen (2007) provides insight into the role of details and imagery in the mutual process of sensemaking. She asserts, "Images, like dialogue, evoke scenes, and understanding is derived from scenes because they are composed of people in relation to each other, doing things that are culturally and personally recognizable and meaningful" (p. 134). When speakers use vicarious proximity appeals, they also strive to create scenes that are recognizable and meaningful; however, their focus is less on people and their actions than it is on place. To make places recognizable and meaningful, speakers present scenes through different strategies that range in focus and vividness. For instance, speaker Dan Bagley says, "the beginnings of a well can be busy, clanking and *dirty, soil dirty, earth dirty* [italics added]." "Dirty" and "clanking" provide dimensions to the vicarious proximity appeal. Some also include odors: "the plant's noise, airborne grit and *stench* [italics added]" or "it [the well site] smells like kerosene and chemicals." Rhetors also populate the scenes they create with life and action: "But *walking up grouse* in Somerset and Mercer and other counties, I have seen that after drilling gas wells *blend with the* land . . . ." Though it is clear that these depictions are not from Keats' "Ode to a Grecian Urn," a highly regarded example of ekphrasis, the speakers do depict scenes with enough specificity to allow their audiences to follow the description and fill in the gaps with their own imagination.

Anti-frackers provide more detailed representations of geographically focused places than do pro-frackers. To support this claim, I consider the specificity of each scene portrayed per appeal as well as the vividness of the included descriptors. In particular, I compare the length of each scene's focus on an individual geographic location. Anti-frackers most often zero in on one distinct and specific location, usually a natural place or well site, and maintain focus on that site for the duration of the appeal. Pro-frackers do just the opposite. They characteristically shift from location to location or provide a broad view of one scene. Where anti-frackers would depict one well site up-close, pro-frackers might depict a large area with multiple well sites. For instance, consider again Loughney's account of fracking operations at the beginning of this chapter next to the following description from Dan Bagley:

> (1) I was raised in Arkansas amid oil wells and gas wells in the natural state, where energy revenues and land management were a boon to wildlife and waterway cleanliness . . . (2) Now, like any industrial construction, the beginnings of a well can be busy, clanking and dirty, soil dirty, earth dirty. (3) But waking up—but walking up grouse in Somerset and Mercer and other counties, I have seen that after drilling, gas wells blend with the land, and then they make money for years.

Bagley's vicarious proximity appeal creates images of three separate scenes. The first is in Arkansas, the second is a generic well site during construction and the third refers to specifically located well sites in two Pennsylvania counties as well as unspecified "other counties." This scene shift is typical of pro-fracker discourse. Pro-frackers also tend to create images that cover broad geographic areas, such as scene (3) where Bagley describes bird hunting for grouse in "Somerset and Mercer and other counties" and refers to no gas well in particular. In contrast to pro-frackers like Bagley, anti-frackers all focus their vicarious proximity appeals on one coherent scene.

To determine which group uses more vivid language, I follow Kaufer and Butler (2000) and turn to Manfred Bierwisch's (1996) work on spatial language. Bierwisch differentiates words that supply spatial information and those that do not. Those that do not are termed (1) aspatial, such as *fear*, *hour*, or *duration*. Other words that do provide spatial characteristics are (2) extrinsically spatial, like "*animal*, *plant*, *tool*, *furniture*," or (3) intrinsically spatial, like "*dog*, *snake*, *child*, *table*, or *pencil*" (p. 51). The difference is that intrinsically spatial words provide an audience with specificity that guides their visualization of the object in space.[3] Extrinsically spatial words "require spatial properties but do not specify them" (p. 52). Bierwisch also adds the (4) strictly spatial, such as "*corner*, *square*, *margin*."

Bierwisch's taxonomy of spatial language is essentially a classification of how easy it is to visualize a word. When he claims that a word like "dog" is intrinsically spatial, but "animal" is extrinsically spatial, he assumes that when hearing the word "dog" a person will have little difficulty imagining that animal in a space (though "a cocker spaniel" would go further in guiding an audience's visualization of that image). When given a word like "animal," however, people require more information to imagine what the writer/

speaker intends to convey. For instance, it is the difference in how closely audiences may come to visualizing the intended scene in the following sentences: "the *dog* ran across the lawn" versus "the *animal* ran across the lawn." Bierwisch's taxonomy of spatial language is useful as a tool for writers who wish to increase the comprehensibility of the scenes they create; as I use this taxonomy, it may also serve as a tool for analysis to compare which speakers present scenes that are easier for their audiences to imagine.

Comparing vividness, anti-frackers use more intrinsically and strictly spatial language than pro-frackers. For example, Bagley, one of the most vivid pro-frackers, still chooses extrinsically over intrinsically spatial language. He uses the extrinsically spatial "wildlife," "waterway," "industrial construction," and "the land." Going against the tendency of pro-frackers to provide sparse description, Bagley does include some intrinsically spatial (as well as auditory) descriptions of the beginning of a well site (clanking and dirty, soil dirty, earth dirty). However, he primarily creates a broad view of multiple scenes using extrinsically spatial language. In contrast, Loughney, an anti-fracker, provides an intrinsically spatial description of one experience at a well site. The images of "containment tanks," "homes," "streams," and "fertile farm soil" add vividness to this scene.

In summary, anti-frackers use more intrinsically spatial description and focus on individual scenes, which they describe in detail. Pro-frackers tend to use more extrinsically spatial description while shifting from scene to scene. When pro-frackers focus on one scene, they provide a "zoomed-out" view of broad geographic areas.

Critical Feature #3: First-Person Reporting

The critical feature (3) *first-person reporting* is an attempt to establish ethos, but it is also used to aid in the visualization of scenes. When speakers recount their own experience, they demonstrate to the audience that they are a credible source of information. This rests on the commonly held assumption that "seeing is believing." To fully establish authenticity, they also use specific, detail-oriented language, which add to one's credibility (Ochs & Capps, 1997). This, in addition to their physical presence in the scenes they depict, makes their role as credible witnesses "discursively verified" (Bucholtz & Hall, 2005, p. 601).

Many of the speakers present their first-hand experience as the only proof anyone should need to support their position. For instance, Ken Gulick says, "I have seen firsthand the site development phase, along with the drilling and completion operations." Gulick advances the fracking scene he has witnessed as a definitive account that can be generalized. In some cases, speakers even attempt to establish a privileged status to their experience. Mike Yilit, an employee of the natural gas industry, states: "I've sat on the iron when they fire the gun to frack pits."[4] He follows this with a sentence that emphasizes his privileged status: "You don't know these things." Both Gulick and Yilit present their first-hand experience as a way to establish ethos and advance deliberative argument.

Time is also an important factor made relevant primarily by claims to recency or long exposure. Thirty-six percent of speakers making vicarious proximity appeals foreground that their first-hand experience has been recent. This increases the authenticity of their claim. Referring to the environmentally conscious actions of drilling companies, Eric Vaccarello says, "I see it every day," while the biologist that Packer quotes says, "We were *just* up in Moshannon" (emphasis mine). Vaccarello not only implies the recency of his time in the place, but also includes support for the implied claim that he has witnessed change by seeing it "every day." Likewise, Packer goes on to quote: "This used to be intact forest habitat. Now it's just patches." This passage both establishes that it is a first-person report and, as I will discuss in the next section, is used to bear witness.

Though speakers like Vaccarello and Packer include references to repeated visitations to the sites they describe, others foreground their long exposure to a place. Long exposure is invoked when speakers explicitly or implicitly claim that they have extensive experience in a particular place, usually as a result of living there. In this hearing, half of the first-person reporting is couched as long exposure. An example comes from St. John, who provides an account of growing up in a Pittsburgh neighborhood. St. John details her life living in close proximity to the Hazelwood Coke Plant, which she claims caused property values to drop. In doing so, she explicitly states, "Let me tell you a similar story from my neighborhood." She implies her continued presence in that neighborhood when she says things like "Most days you could smell the sulfur. . . ." Statements like this present her account as definitive of life in that particular time and place. In citing long exposure, speakers justify their inclusion in the hearing and build credibility.

The specificity used to describe the places that they claim to have inhabited varies between pro- and anti-frackers. All but one of the six anti-fracking speakers locate their first-hand experience in one distinct geospatial place, such as a Pittsburgh neighborhood from St. John or a well site from Packer and Loughney. As mentioned, pro-fracking speakers situate their experience much more broadly, either covering larger geospatial areas, such as "down there" in Washington and Butler counties, like Valentas states, or in multiple sites, like Yilit describes.

Juxtaposed to the first-person narratives, two anti-fracking speakers present accounts that are not their own. Mel Packer exemplifies this type of appeal by saying, "a plant biologist I know e-mailed me the following." Meanwhile,

Jeanne McMullen reads "excerpts of an e-mail from a resident in Hickory." Buttny and Cohen's (2007) finding that reported speech is used in public hearings on environmental issues to formulate risk may in part explain why reported speech is only invoked by anti-frackers (p. 753). Since they may have less access to fracking sites, in some cases anti-frackers depend on reported speech to present a scene that can be imagined by the audience with the rhetorical purpose of conveying a sense of risk.

Critical Feature #4: Bearing Witness

The final critical feature of vicarious proximity appeals, bearing witness, introduces the speaker's external evaluation. That is, an evaluative voice that "seeks to convince hearers or readers by providing interpretations in the storyteller's voice, from outside the story" (Tannen, 2007, p. 136). Tannen, citing Labov (1972), explains that internal evaluation is more persuasive than external (p. 136); however, all vicarious proximity appeals include explicitly external evaluation, which may be a function of the rhetorical situation of the public hearing where explicit evaluation is expected.

The basis for this external evaluation is the speaker's experience with the transformation of a place. The term "bearing witness" is often associated with atrocities or miracles. The participants in the hearing do not go to such extremes; however, they do present strong positive or negative evaluations. By offering vicarious experiences of places, they both present what they intend to be taken as objective truth and as a moral imperative based on that truth.

As an identifiable feature of vicarious proximity appeals, to bear witness speakers tend to use visually vivid depiction and attach an abstract concept that either implicitly or explicitly advances an evaluation. This can be done quite simply. For instance, Patrick Imbrogno says:

> I grew up in the middle of a national forest, probably one of the most environmentally sensitive areas of the state. Native brook trout. I've seen them co-exist with a lot of these—Just be reasonable. Work with people. Come up with a solution, you know, some kind of joint effort with the industry to make this happen.

Imbrogno demonstrates that his account is a first-person report by mentioning his experience growing up in a national forest. He then says, "native brook trout," which I consider to be (2) visual vividness in that he populates the place he has specified (the national forest) with an entity that the audience can imagine. This becomes bearing witness when he says that these trout "co-exist" with what we can assume to be fracking operations. The abstract concept of co-existence is to be taken as a positive evaluation of the relationship between fragile nature and energy extraction. To convey that entities and/or processes "co-exist," speakers imply that they have observed a site over an extended period of time. In this manner, Imbrogno bears witness to the transformation of a familiar place and evaluates the hearing by drawing a comparison.

Critical Features and Variables in Concert

Now that I have shown how each critical feature of a vicarious proximity appeal functions, I turn to the nonessential, variable components of these appeals. In doing this, I will provide full excerpts to not only demonstrate the ways variables are employed, but also to show how they interact with other elements of a full vicarious proximity appeal.

The four variable components of vicarious proximity appeals are: (a) *heritage*, (b) *common place*, (c) *analogous situation*, and (d) *invitation to visualize*. Except (d) *invitation to visualize*, these variables can be considered supplementary aims to the main rhetorical effect of a vicarious proximity appeal.[5] For instance, heritage is commonly invoked as an additional layer of importance to a place. Similarly, (b) *common place* attempts to change the subject position of the audience to embrace their responsibility for the future perception of their community. By using variable (c) *analogous situation*, speakers present a vicarious proximity appeal that does not directly deal with the subject of the hearing, but relates closely enough to provide historical context. Finally, (d) *invitation to visualize* is a discourse move speakers use to explicitly prime the audience's imagination.

The first vicarious proximity appeal in the hearing, by Mel Packer, provides an example of three of the four variables in use, all but (c) *analogous situation*. Packer states:

> A plant biologist I know e-mailed me the following. We were just up in Moshannon State Forest in Clearfield County. *You can't imagine the damage that's done.* There's trucks every five minutes on the forest roads that were formerly small and quiet. They've cleared huge patches of land for well pads. The lights and noise are constant when actively drilling. This use to be intact forest habitat. Now it's patches.
>
> As one driller said to us, these whole woods are going to be tore up by the time we're done here. *It's not Commonwealth land anymore.* The local people can't hike, hunt and fish when the forest is transformed in this way, and the wild resources. . . . *This should be the heritage for future generations.* It's sabotaged for short-term profit [italics added].

Reading the text from beginning to end, the first variable we encounter is (d) *invitation to visualize* (in italics). Even though the email Packer quotes reads "You can't imagine . . . ," I take this as a clear invitation to visualize the scene that he is about to present. This instance functions as a challenge to the audience to push their imagination to visualize the devastation Packer describes. Other speakers explicitly and directly invite their audience to imagine: ". . . close your eyes for a minute and imagine. . . ." In all instances, (d) *invitation to visualize* functions in concert with (2) *visual vividness* as speakers prompt their audience to prepare for the depiction to come. Variable (d) is also presented with (1) *focus on specific place through place-names and space-relational descriptors* as rhetors situate the scene within a geospatially defined context.

By invoking the land that belongs to those who identify with the Commonwealth of Pennsylvania, this speaker also makes an appeal to the (b) *common good.* A common feature of (b) *common good* is that speakers define what activities constitute appropriate use of land. For Packer's biologist, appropriate use of Commonwealth land is for recreational activities. He also includes "wild resources," which must be taken to be distinct from resources like natural gas. While I call this span of text (b) *common good* due to Packer's illumination of common ownership of land, at the same time Packer is developing (4) *bearing witness.* By normatively ascribing value to Commonwealth land and delimiting appropriate use, Packer sets up his evaluation of what happens "when the forest is transformed in this way." In this sense, this passage represents both (4) *bearing witness* and (b) *common good.*

Packer's speech also contains (a) *heritage* in the concluding lines of the excerpt. He states that the common land he has established should be considered the heritage of future generations. This puts the audience in a temporal relationship with the land described. In doing so, (4) *bearing witness* is again a part of the equation. Indeed, (4) and (a) *heritage* go hand-in-hand, as speakers evaluate both. For instance, pro-fracker Dan Bagley includes the future by invoking the profitability of wells over time in conjunction with an implied evaluation of this process in terms of profitability and its negligible impact on the environment when he states: "I have seen that after drilling, gas wells blend with the land and then they make money for years." Bagley does not explicitly state that the profitability of wells is a form of heritage, but he implies it in the temporal phrase "for years," which suggests that earnings will be enjoyed well into the future.

Finally, several of the vicarious proximity appeals in this public hearing make use of an (c) *analogous situation,* which contains all four critical features but is not focused on an experience with natural gas extraction in Pennsylvania. I have previously mentioned Kate St. John's account of the Hazelwood Coke Plant, which is a strong example of (c). For St. John, a polluting agent in close proximity to residential areas is analogous. For someone like Charles Clark, the analogous situation is not as apparent:

> I grew up in Pittsburgh. When I was young, the community street lights were on. Every morning when you'd wake up, there'd be a half an inch of chemical dust on your car. We got over that. I had asthma as a child. I don't have asthma anymore. The air in Pittsburgh is good. The water is good. I can bass fish at the Point. I couldn't do that 40 years ago. The laws that we have in this state protect the environment, and they're being enforced. They will take care of this industry, but we need the energy now. So as far as I'm concerned, it's drill, baby, drill.

The analogous situation that Clark sets up is predicated on his assumption that hydraulic fracturing will be no different from other forms of environmental pollution that Pittsburgh has faced in the past (largely from the coal and steel industries). Clark fulfills each of the conditions of a vicarious proximity appeal by (2) providing a series of images (1) situated in Pittsburgh (4) that juxtapose the past with the present to advance an evaluation (3) based on what he has seen. Typical of pro-frackers, Clark quickly jumps from scene to scene. The connection between these scenes is the improvement of the environment due to the positive influence of regulation and the nullifying effect of time.

VICARIOUS PROXIMITY AS A PRACTICE OF ORDINARY DEMOCRACY

I started this chapter with the question, "How is a sense of proximity created in discourse?" I have answered this question by uncovering the four critical features and four variables of vicarious proximity. It can now be said with certainty that as a place-based appeal used in an environmental issue, vicarious proximity is not a tactic particular to environmentalists; it is practiced nearly equally by participants on both sides of the issue. Vicarious proximity should not be thought of as a specific kind of *environmentalist* rhetoric. Instead, vicarious proximity appeals are a ubiquitous rhetorical resource of ordinary democracy in disputes over land use. Indeed, Scarff's (2021) application of vicarious proximity to the Mountain Valley Pipeline issue in West Virginia and Virginia offers support to the notion this concept has broad application.

The remaining two questions that opened this chapter are the focus of this closing section: What does the vicarious experience of proximity achieve in a site of ordinary democracy? How can we put this knowledge to use in future advocacy efforts? We have seen that though the same critical features and variables are used to create vicarious proximity appeals, there is a clear difference between the ways people who support and oppose hydraulic fracturing construct these appeals. Anti-frackers position the audience on the ground; as if they were standing at a fracking site, while pro-frackers tend to provide their audience with an aerial view of larger swaths of land. This disparity amounts to a struggle for control over the representation of place. Anti-frackers assert more control over the imagery they depict by providing more specific representations of place. This choice may be more likely to fail to connect with their audience if the audience is unfamiliar with the specific place mentioned or if the depiction does not correspond with what the audience has experienced. Tannen (2007) explains that details create involvement with the audience, but this can be overdone. She states, "When one feels that a written work of art or persuasion is demanding too much involvement—trying to pull at one's 'heart strings'—one resists . . . and labels that work manipulative or sentimental" (p. 159). However, if the audience is familiar with the place and accepts the description, they may be more likely to agree with the evaluation. In this, it may well be the more descriptive and specific these speakers are, the higher the risk of the appeal.

Pro-frackers, who offer less descriptive accounts that cover broader areas, allow for audience participation in the construction of the scene. If the audience wishes to engage with the scene, they draw from their own experience to enter into a proximate relationship with the depicted place. This "aerial view" may be more likely to succeed if an audience is equipped to imagine the scene depicted, but it may also fail if the scene is too vague to connect to any particular place in a person's experience. Even though this less descriptive version of the appeal is not as risky, it is more likely to be dismissed and forgotten. It may be no coincidence that the two instances of speakers presenting the vicarious proximity appeals of outsiders to the hearing (in the form of e-mails that were read to the audience) belonged to those on the opposition.

As Kalin and Frith (2016) write, depictions of place that come across as "striking and singular," which are more common among anti-frackers, are more likely to be remembered (p. 227). The close-up views anti-frackers offer are meant to provoke an emotional, fear-based reaction that audiences do not forget. Pro-frackers, as we have seen, describe places in more general terms to assuage the audience's doubts and fears about the fracking process. Once described, the places can be forgotten and people are left with the feeling that "everything is fine." The distinction between the ways the two sides utilize rhetoric to describe place shows how the hearing becomes not just a competition over a place, but over memory as well. Through a process of enthymematic reasoning, "rhetoric crafts and utilizes forms of public remembrance in the service of disciplining memories and shaping [*sic*] a framework in which the experience of the past is cultivated and contained" (Phillips, 2010, p. 217). In this manner, the landscape of a community becomes inextricably linked with the politics of the public: "communities conceive, maintain, or contest collective remembrances in accord with sociopolitical interests" (Vivian, 2018, p. 292).

Although proximity is used in discourse to negotiate the meaning of place in a community, it is very rarely an "eco-centric" appeal, or an appeal that characterizes an entire eco-system as equally important (Killingsworth & Palmer, 1992). The value of the land has the same value as the people. On the other hand, vicarious proximity appeals are almost exclusively "homo-centric" in that they are focused solely on impact to and from humans. This may be counter-intuitive for those who oppose hydraulic fracturing. Environmentalists are often thought to value the whole ecology over human advancement, but what I have found is that even when the devastation of a forest is the focus of the appeal, the appeal is linked to the human impact of this land-based intervention. As a constraint of democratic practice, this homo-centric focus can be understood as a way to argue for the common good. Though it could be said that the soundness of the ecological system does involve the common good, the participants in the public hearing deal with common good as it relates to the (a) *heritage* that will be left for future generations and to the characterization of the place as a (b) *common place* as it is used and shared by people in the community. This is further supported by the (1) *specific place references* that accompany each appeal, as recognizable communal entities are presented as at stake. Even in cases where participants focus on their privately-owned land, this is characterized as representative of what would happen in the more clearly shared spaces of public life. In sum, vicarious proximity is used regardless of ideological affiliation to argue for the common good. Such knowledge can be useful to those attempting to shape public discourse.

Vicarious proximity appeals also help participants in a public hearing negotiate constraints of democratic practice. These appeals help an individual present themselves in relationship to their situated knowledge. This primarily takes the form of both a (1) *focus on specific place through place-names and space-relational descriptors*, and (3) *first-person reporting*. These critical features establish the ethos of the participant as someone who can be trusted; as they have witnessed the scene they describe and provide a clear context for the action either within the confines of Allegheny County or in a place that is deemed relevant. Alongside industry experts, scientists, professional activists and others whose reputation precedes them or who more easily construct their ethos by citing their professional affiliation, participants who use vicarious proximity appeals enter into the public realm via their relevant experience in place. They present themselves as experts on the land that is to be affected or that already has been affected. This is further bolstered by their use of (2) *visual vividness* to create what can be assumed to be a realistic portrayal of what they have seen. Vicarious proximity appeals establish ethos in a public setting that privileges the discourse of those with more institutionally recognized perspectives and builds a relatable ethos appeal for those with institutional credentials.

This view aligns with Iris Marion Young's (1996) "communicative democracy," which privileges (1) storytelling, (2) rhetoric, and (3) greeting. Though (3) greeting is not prevalent in these hearings as part of vicarious proximity appeals, all participants do greet their audience, state their name and where they are from. The important elements of communicative democracy that are exhibited through vicarious proximity appeals are the first two of these features. Vicarious proximity appeals are clearly narratives that often do involve rhetoric as Young defines it (associated more with ethos and pathos than logos). So, this analysis shows that the public hearing operates as a site where some participants choose to tell their own story. In this way, they advance their interpretation of events and, ultimately, bear witness. What vicarious proximity adds to this account is the role of place and rhetorical vision in this process (Bormann, 1972). Storytelling is used to negotiate the meaning of place in a community and, by doing this, establish a shared rhetorical vision of the satisfactory result of deliberation. Of course, this aim is not achieved in the setting of the public hearing, as no decisions are made and there is no institutionalized interaction between participants. Despite this, because vicarious proximity appeals are simple narratives with a well-defined setting, few characters, and a tendency to exhibit action that is not very complex, they are easily remembered. This increases the chances these appeals will circulate, as evidenced by the two instances of speakers recounting the stories of others.

What this adds to the understanding of ordinary democracy is that in environmental disputes such as this one, personal experiences with place are not only a source of conviction; they are also a resource to be drawn from, shared and experienced vicariously by others in the advancement of deliberative argument.

CONCLUSION

In this chapter, I have shown that proximity is rhetorically constructed by rhetors and audiences through vicarious proximity appeals. Vicarious proximity appeals open an intersubjective space where the meaning and value of place can be negotiated between a rhetor who advances a position on an issue by bearing witness and an audience who co-constructs the depicted scene and either accepts or rejects the embedded argument. This task is accomplished by presenting a descriptive account of a place that is couched in a person's first-hand experience and passes judgment on what they have witnessed in that place. Through the resulting analysis, I show that vicarious proximity is used nearly equally by those who support and oppose hydraulic fracturing.

This appeal is used to establish ethos and authority in public settings where technical expertise typically wins the day. Vicarious proximity appeals are an alternate way of presenting situated knowledge under such circumstances. In the next chapter, I continue to investigate the role of vicarious proximity appeals in public rhetoric, broadening our focus to include additional public hearings from other places deliberating over fracking.

NOTES

1. I preface this section with the disclaimer that though John Muir was a nature writer and an environmentalist, vicarious proximity appeals are not the sole purview of environmentalists. As we will see in the analysis of the public hearing, these appeals are used equally by those who support and oppose hydraulic fracturing.

2. It is important to note that not all vicarious proximity appeals depend on the first-hand experience of the speaker themselves. Though it may add some degree of confusion to the term "first-hand experience," one of the virtues of vicarious proximity appeals is that, by being narratives, they are easily remembered and reproduced by those who have heard them. Because of this, others may represent vicarious proximity appeals of which they have no first-hand experience. Nevertheless, vicarious proximity appeals do depend on someone's first-hand experience, whether it is reported by the person who has experienced it or represented second-hand.

3. Bierwisch admits that within the intrinsically spatial there are degrees such as *bird* and *parrot*.

4. Yilit is an outlier to the aerial perspective that is common among pro-frackers. However, Yilit does jump from scene to scene in his account, relocating his audience's focus from his family's farm to his job site "sitting on the iron."

5. All vicarious proximity appeals present a scene for the audience to visualize. The variable *invitation to visualize* emphasizes the invitation and is reserved for instances where speakers explicitly prompt an audience's imagination. This is accomplished by using an imperative statement (e.g., "Close your eyes and imagine . . .") or with a prompt that signals the beginning of a narrative (e.g., "I'm going to tell you a story about my neighborhood").

Chapter 4

Vicarious Proximity as a Micro-Rhetorical Strategy of Image Repair

Pro- and anti-frackers ground their arguments in the land using vicarious proximity appeals. Through place-based arguments, they share value hierarchies with their communities and those making decisions about their futures. As shown in the previous chapter, they incorporate vicarious proximity appeals at similar rates. However, the narratives of place that they tell represent scenes differently: pro-frackers provide aerial views of the land from a distance while anti-frackers bring their audience up close. For both sides, place functions as a storehouse of memory, which means that these differing representations greatly impact the issue at hand. Depending on the rhetorical technique used, the audience might remember a vast, bucolic countryside with small pockets of industry or a favorite forest trail despoiled by leaking machinery.

These striking disparities in place representation can teach us how ideological disputes such as this one represent different ways of seeing the world. What else are people communicating through verbal representations of their environment? In this chapter, I continue to explore this question by investigating the continued role of vicarious proximity appeals in environmental rhetoric. What I am looking for here are not the overt arguments for or against hydraulic fracturing, but the nuances of place representation that differentiate the worldviews of supporters and the opposition. To do this, I expand the corpus to include seven different public hearings held in three states from 2008 to 2014. I apply both computational and discourse analytic methods to compare the rhetoric of these groups. The result is a dive into the micro-rhetorical.

My major finding is that in addition to its usefulness in advancing deliberative argument, vicarious proximity serves a micro-rhetorical function as *apologia* or image repair. What I mean by this is that speakers in hearings construct places in ways that are designed to counter stereotypes about the groups they represent. How they achieve this is detectable through a combination of discourse analysis and computer-aided rhetorical analysis.

A computer-aided approach to rhetorical analysis helps reveal minute details in discourse that would otherwise be nearly impossible to discern. This chapter offers a new look at apologia as well as a blended approach to rhetorical analysis using the close focus of discourse analysis alongside the larger-scale corpus view afforded by computer-aided textual analysis. This offers digital humanities scholars a widely applicable analytical approach. Broadening our understanding of apologia may facilitate the methodology of rhetorical analysts, allowing them to sift through more text than ever before, which often now takes the form of short comments, captions, and Tweets. From the micro-rhetorical perspective, these short exchanges suit our purposes because our interest is not in long bouts of sustained argument, but in the implicature of representation that can be gleaned from a few words.

Public hearing speeches are most often held to three minutes each and characterized by their length and density. Speakers pack a lot into their short time before the council; to meet the exigence of the situation, they make their position explicit. Yet, insults do fly and tempers do flare despite the planned discourse most bring to the podium. Interspersed throughout their arguments, pro- and anti-frackers attempt to discredit one another. These groups do not simply argue why they are for or against fracking to the governing bodies seated at the front of the room; they also attempt to explain why the other side has it wrong.

Our current political climate is characterized by adversarial relationships across ideological lines. Not just politicians, but friends and family, too, must defend themselves against attacks made by those who are unable and unwilling to see from our perspectives. Moderates are harder to find as our current world forces us to choose sides. How can we effectively defend against stereotypes and unfair accusations, especially when the situation does not allow for sustained response? Speakers in these seven public hearings use their place-based arguments to advance rhetorical defense against stereotypes, but they do not explicitly address these claims. Their defense is located in the realm of the micro-rhetorical. In particular, anti-frackers, associated with environmentalists, are often characterized by pro-frackers as idealistic, misguided, and more concerned with fervent anti-industrialism than with facts. Aubrey McClendon, CEO of Chesapeake Energy, draws a line between "fracktivists," who live in a "fantasy land of a world without fossil fuels," and his people, whom he dubs "factivists" (Phillips, 2011).

This same message appears in public hearings, as anti-frackers are deemed out of touch with reality. One speaker in particular in the March 2011 Allegheny County hearing that has been analyzed in this book exemplifies this position to the council:

> I'm here to ask this Council not to be bullied by zealots asking you to believe the worst that can happen will be the average. These zealots come from the same mold that told our great-grandfathers, the George Westinghouses, alternating power and light was much too dangerous and that people would be electrocuted daily in the streets.

In other words, anti-frackers are Luddites, cut from the same cloth as those who have opposed progress for generations. What drives them is an ideology based on an irrational fear of technology. Consistent between McClendon's and Howard's messages is the implication that anti-frackers lack any empirical foundation to their claims.

Meanwhile, anti-frackers also criticize their opposition, characterizing pro-frackers as out of touch with reality and lacking awareness or appropriate consideration of the risks involved in natural gas development. The risks to environment and wildlife are placed on the families of local community members. Pro-frackers are maligned as lacking foresight and blinded by greed for short-term profit. Bernie Eckley, an Allegheny County resident, explains his view in relationship to the gas industry and its supporters:

> We don't need the natural gas. We do not need to endanger generations for money. That is what it's all about. People, animals, more jobs. None of that matters. Eventually, we are killing our children and causing disease and sickness.

Anti-frackers tend to imply the shortcomings of pro-frackers in statements like Eckley's, which identifies anyone who opposes his claim as complicit in the death of children and the spread of preventable disease. Eckley also foregrounds the "generations" that may be affected by rampant gas extraction and implies that pro-frackers do not care about the future.

Activist filmmaker Josh Fox, director of *Gasland* (2010), who testified at a 2008 New York State hearing, supports this view. He explains his cross-country trip to various gas extraction sites, where he interviewed affected people to identify public health risks. As a secondary aim of his speech, Fox helps to create this pro-fracker identity by emphasizing the disconnect between pro-fracker rhetoric and what he saw:

> I want to stress again every single person that I talked to had heard this line . . . that it's not going to be that impact, that it won't hurt your land, it won't hurt your health, you're going to make a lot of money. What I heard was absolutely devastating health effects on populated areas, health effects on forests, health effects on water.

Beneath it all, anti-frackers characterize pro-frackers as uncaring and simply driven by thoughts of financial gain. They strive for unfettered gas industry expansion in the face of what anti-frackers see as obviously unacceptable risks to people and place.

In response to this vitriol, I have found that one way that speakers in public hearings use vicarious proximity appeals is to go on the defensive and repair their side's image. In doing so, they participate in a rhetoric of self-defense, known as *apologia.* By engaging in apologetic discourse and defending their side, speakers make discourse moves that potentially build common ground between participants with different ideological positions, complicating the seemingly black and white view each side has of the other. Speakers nuance the generalizations that the other side uses to label them. For instance, if characterized as driven solely by economics, "debaters may seek shelter from attack by claiming for themselves a breadth of attitudinal experience denied them by their opponents" (Killingsworth & Palmer, 1992, p. 12). The attitudinal experience that Killingsworth and Palmer describe manifests strongly in place-based experiences in Marcellus Shale public hearings.

My analysis shows that vicarious proximity appeals are used for image repair in three distinct ways: (1) Pro- and anti-frackers respectively insert or remove themselves from vicarious proximity appeals, creating *a more subjective or objective representation of place.* (2) Anti-frackers also make use of *scientific discourse* to align themselves with *scientific rationality* and counter views that they are emotionally driven and disconnected from reality. (3) Pro-frackers attempt to counter the view that they have no concern for nature or the future of the places from which they would like to extract gas by characterizing themselves as stewards of the environment and emphasizing their heritage. In short speeches of three minutes in length, citizens explicitly argue for their position on hydraulic fracturing, but they also craft scenes in ways that resist dominant narratives by incorporating sophisticated strategies at the micro-rhetorical level.

APOLOGIA AND IMAGE REPAIR

Since the beginning of the televised news media, people who have done wrong have often chosen to utilize rhetoric to defend their actions and repair their public image. From Nixon's Watergate to Bill Cosby and Harvey Weinstein, there is no shortage of public exchanges on news or now social media that reconstruct the identity of a person who has been publicly shamed or humiliated. Rhetorical scholars, who study apologia, or verbal self-defense, can find their studies in a wide range of historical examples. The groundwork of their scholarship could be pinpointed to Ware and Linkugel's (1973) "They Spoke in Defense of Themselves." Communication scholar William Benoit has done similar (and more widely known) work on "image repair theory." Since then, apologia and image repair have served as complementary theories of rhetorical defense that typically examine powerful figures with political or celebrity status (or both). Corporations, too, engage in image repair with a host of rhetorical strategies not available to the individual, including scapegoating (Hearit, 1995, p. 13).

Apologia is a paragon of the idea that genres live and breathe in cadence with changes in society. From the time Aristotle introduced apologia as a counterpart to forensic *kategoria* [accusation] (1358b), the form of this genre of rhetoric has remained quite solid, identified by specific situational markers (e.g., accusation precedes apologia; Downey, 1993, p. 58). Apologia used to include a response to a public attack on a person's character after a perceived wrongdoing (Coombs et al., 2010, 339). Ware and Linkugel explain:

> In life, an attack upon a person's character, upon his worth as a human being, does seem to demand a direct response. The questioning of a man's *moral nature*, *motives*, or *reputation* is qualitatively different from the challenging of his policies. Witnesses to such a personal charge seem completely and most easily satisfied only by the most personal of responses by the accused. In the case of men and women of position, this response is usually a public speech of self-defense, the apology. (Ware & Linkugel, 1973, p. 274)

Apologia required a specific exchange: one person clearly and publicly accusing a wrongdoer. The accused, also an individual, rose to their own defense in a court of law. Recent societal changes and advances in rhetorical methodology, however, have led to a re-examination of this concept, which has increasing relevance in our lives today. Apologia now allows corporations (Hearit, 1995; Hearit, 2018; Riley 2015; Marsh 2005; Coombs et al. 2010) and places (Prasch 2015; Avraham 2014) as to act as individuals and third parties to serve as defenders of the accused (Benoit et al., 2015; Benoit, 2017). Direct accusation is no longer the required initiator of apologia as "negative messages, stereotypes, perceptions and generalizations spread or reported upon" may also put us into this defensive posture (Avraham 2014, p. 225). And, perhaps most importantly, scholars now recognize that apologia may serve broader purposes beyond mere defense.

We now know that apologia may serve dual motives, working both on the defensive and offensive (Downey, 1993). Downey situates this shift as a response to contemporary times, where the venue of apologia is less often the courtroom and more often the newsroom. Those accused *elect* to defend themselves rather than be forced into defense as part of judicial proceedings (p. 53). Not only have the available participants broadened, but the venues for apologia have too. These changes have allowed rhetorical scholars (and scholars of communication) to examine disparate discourses through the lens of apologia, leading to insights about the nature of public defense.

Study of apologia as a distinct genre of rhetoric began with an interdisciplinary approach that drew from psychological research on conflict resolution. As mentioned, Ware and Linkugel (1973) introduced apologia to many rhetorical scholars with their seminal generic criticism, "They Spoke in Defense of Themselves." Through genre criticism and theories of conflict resolution, they defined different stances and postures of apologia that range from denial to bolstering and transcendence. In their study, Ware and Linkugel explain the motivations and limits of each. Their approach particularly describes the apologetic rhetoric of political figures. Ware and Linkugel themselves provide examples of Ted Kennedy's "Chappaquiddick" speech, trial rhetoric from Eugene Debs and Clarence Darrow, as well as Richard Nixon's "Checkers" speech.

Apologia is much like the related concept of "image repair," which was extensively studied by William L. Benoit (1995, 1997, 2000) and Benoit and Drew (1997). Benoit, along with Ware and Linkugel, examine cases where public figures or corporate entities respond to criticism of wrongdoing. Kennedy's "Chappaquiddick" speech, in which he sought absolution from his involvement in the death of Mary Jo Kopechne, is a frequent example of this type of rhetoric and highlights apologia's role in defending one's bad behavior. In contrast, John Henry Newman's (1864) *Apologia Pro Vita Sua* is a noted example of apologia that defends one's deeply held beliefs. Although neither Ware and Linkugel nor Benoit not address apologia used like Newman's response to criticism of Catholicism, apologetic discourse in hydraulic fracturing hearings is more closely aligned with Newman's use. Speakers more clearly defend their values than their behavior.

Genre criticism of apologia and image repair still hold a durable sequence of accusation to defense, even though this genre has shifted to allow for a wider range of participants. Direct accusation remains a defining feature that is explicitly addressed in scholarship on apologia. However, in public hearings apologia manifests not as explicit defense, but instead as a micro-rhetorical sub-aim of deliberative discourse. Contrary to typical cases of apologia, this different form of apologia that I identify in public hearings is implicit rather than explicit and is enacted in discourse through syntactical and lexical choices that oppose stereotypes. For instance, when creating depictions of place, anti-frackers remove themselves from these scenes to advance what seems an objective account of observable phenomena. I argue that this seemingly objective representation serves as an implicit response to criticism that this group is out of touch with reality and that their subjective fears overcome their faculties of reason.

The features of apologia apply not only to public figures and corporations that protect their image through mass media discourse, but can also be enacted through less explicit means as a sub-aim of larger rhetorical appeals. By specifically analyzing vicarious proximity appeals using a corpus-based approach, significant differences between the discourses of pro- and anti-frackers reveal attempts at image repair that may otherwise go unnoticed. Nevertheless, if we accept that "every choice is strategic, in the sense that every utterance has an epistemological agenda" (Johnstone, 2011, p. 54), these different approaches to creating vicarious proximity appeals become meaningful. Implicit responses to stereotyping that attempt to bolster the audience's perception transcend unfair accusations and suggest that corrective action may complicate the black and white image participants in the Marcellus Shale public have of one another. This may provide some hope of finding common ground between two groups in opposition.

CORPUS APPROACH AND COMPUTER-AIDED RHETORICAL ANALYSIS

Seeking to better understand the difference between the ways pro- and anti-frackers invoke vicarious proximity, I have increased the size of my corpus from the previous chapter and employ computational methods to guide a close reading of it. The texts that make up this corpus are the vicarious proximity appeals exhibiting the four critical features established in the previous chapter: (1) *focus on specific place through place-names and space-relational descriptors*, (2) *visual vividness*, (3) *first-person reporting*, and (4) *bearing witness*. They were taken from seven public hearings on hydraulic fracturing from 2008 to 2014 located in the states of Pennsylvania, New York, and Illinois.

The resulting corpus contains an even distribution of vicarious proximity appeals between anti- and pro-frackers. This corpus contains 30 anti- and 30 pro-fracker vicarious appeals. The pro-fracker appeals total 5,090 words while the anti-fracker appeals total 5,599. As I am interested in the specific role that vicarious proximity appeals play in these hearings, I have selected only vicarious proximity appeals for this corpus and have cut all other segments of these longer speeches.

The computer-aided component of this analysis begins with the corpus analysis program, DocuScope. DocuScope scans through texts based on dictionaries containing 735 million language patterns that are categorized as part of more than 25 clusters of language use, more specific dimensions, or the most specific level of over 100 "Language Action Types" (Ishizaki & Kaufer, 2011, p. 256). Kaufer and Ishizaki, the creators of the DocuScope, state that the program was created so that "aggregate features of micro-rhetoric could be interpreted as macro-rhetorical strategies underlying the design of a single

Table 4.1 Dates, Locations, and Titles of Public Hearings Investigated

Date	*Location*	*Hearing Title*
November 12, 2008	Bath, NY—Haverling Central School	Public scoping meeting for supplemental generic environmental impact statement on DEC's oil and gas regulatory program for the Marcellus Shale.
December 4, 2008	Loch Sheldrake, NY—Sullivan County Community College	Supplemental generic environmental impact statement draft scoping meeting on DEC's oil and gas regulatory program for the Marcellus Shale.
July 21, 2010	Pittsburgh, PA—Allegheny County Courthouse	Marcellus Shale Public Hearing
March 10, 2011	Pittsburgh, PA—Allegheny County Courthouse	Marcellus Shale Public Hearing
February 7, 2013	Moon Township, PA—Robert Morris University	Bill No. 7387-13: Pittsburgh International Airport Natural Gas Extraction
December 16, 2013	Effingham, IL—Holiday Inn Hotel	Public hearing proposed administrative rules for hydraulic fracturing.
April 15, 2014	Pittsburgh, PA—Allegheny County Courthouse	Proposed lease for oil and gas rights underneath Deer Lakes Park

text or a collection of texts" (p. 256). This program is particularly useful when dealing with a corpus of texts, as in this study, by providing the rhetorical analyst with a precise view of recurring textual features that a serial reader would have a very difficult time identifying. Also, this program is able to quantify results, which can be used to perform factor analysis. A statistical analysis suite was also used to discover "communicative function(s) most widely shared by the features grouped on each factor" (Biber, 1986, p. 391). Computational analysis resulted in two significant findings that I have chosen to explain and elaborate in this chapter.

Computational analysis through DocuScope allows an analyst to focus attention on varying levels of specificity within a corpus. This provides insight into macro- and micro-features of discourse that can be proved statistically significant determiners of difference between corpora. At the top level of analysis, the "cluster" level, DocuScope (v. 10-26-2014) quantifies language use into 25 broad categories called clusters. An analyst is also able to look at more specific features of language divided into 159 "Language Action Types" (LAT's). For instance, the *Descriptive* language cluster is activated when language "conjures in the reader's mind concrete spaces, scenes, objects with lively and colorful properties, human speech, and dialog" (DocuScope). This larger cluster is divided into LAT's of the following: *Sense Property*, *Sense Object*, *Space Relation*, *Scene Shift*, *Motions*, *Comparison*, and *Resemblances*. My findings are based on results from both the broad cluster-level analysis and the much more fine-grained though selective LAT analysis.

When interpreting a factor that results in significant variance between corpora, it is useful to think of each variable (cluster-type or LAT) as a switch. Analysis of variance considers which combinations of on or off switches co-occur. For instance, when the pro-fracking corpus has the *Relations_Positive* switch on, analysis of variance might show (as it does in my corpus) that it is reasonable to expect that the *Emotion_Negative* switch is off and vice versa for the anti-fracking corpus. This becomes more interesting when multiple variables co-occur in concert in conjunction with the absence of variables. These co-occurring and absent variables can provide an analyst with insight that focuses attention on salient differentiators that may not be otherwise apparent.

However, computational analysis does not allow a rhetorical analyst to forgo close reading. Rather, computationally derived results are most useful when used as a lens for subsequent analysis of the text. Additionally, familiarity with one's corpus is paramount in order to understand the significance of computational results. In my case, close reading allowed me to focus on the ways in which each of the four critical features of vicarious proximity are actualized in each speech. Close reading also provides a way to "eye-test" computational results to determine whether such findings are readily apprehended by a serial reader. Through close reading, I have determined that though some results are unexpected—intuition suggests pro-frackers would be less sentimental than anti-frackers due to the very stereotypes they push against—these computational findings are borne out through close attention to the texts.

FINDINGS: STEREOTYPES AND MICRO-RHETORICAL RESISTANCE

Conforming to stereotypes about each group, there are significant differences between vicarious proximity appeals made by pro- and anti-frackers at the cluster level. What I see as most interesting about this is that viewed broadly, pro- and anti-frackers use vicarious proximity in expected ways; but, as I will show, a finer-grained analysis results in more surprising use of vicarious proximity that functions as a rhetorical counter-strategy to negative stereotypes. In other words, vicarious proximity is shown to be a strategy of image repair. I will focus first on the ways speakers conform to stereotypes before I discuss vicarious proximity as image repair.

Both sides are maligned by the other for their perceived weaknesses. Pro-frackers are stereotypically represented as hard-hearted, lacking compassion for the land and its people and downplaying the risk of hydraulic fracturing. Anti-frackers, those aligned with environmentalism, are stereotypically attacked for lacking scientific grounding, their supposed naiveté about the economics of the situation, and exaggerating risk. Comparative analysis of the vicarious proximity appeals from each group shows that these stereotypes are, in part, a result of rhetorical choices. Pro-frackers may be maligned for lacking heart due to their tendency to exhibit high levels of what Kaufer and Ishizaki term "Academic Language," which is composed of words that occur at low frequencies in standard English language corpora like the Frieburg-Brown Corpus of American English. This is often accomplished by the use of abstract or specialized vocabulary. It is important to reiterate that my corpus consists of only vicarious proximity appeals, which means that this abstract and specialized language is found in language used to create an experience of place for its audience.

For pro-frackers, defining place provides an opportunity to demonstrate specialized knowledge of abstract processes. For instance, one speaker says, "*Economic vitality* cannot exist in an area where job and *population*—with job and *population losses*" (emphasis added). Within her vicarious proximity appeal focused on describing an economically depressed Ohio mill town that has rebounded since the discovery of natural gas, Puzzuole interjects this claim about economic vitality, an abstract measure requiring specialized knowledge.

Another example of the focus on abstract concepts as motivator of action comes from DeIuliis. He states, "In no other part of the country and maybe no other part of the world, can lay claim to the energy position that we enjoy here in southwestern PA." Like Puzzuole, DeIuliis motivates support based on an abstract concept (energy position). This is representative of pro-frackers' tendency toward abstract concepts as incentives to proceed with gas exploration. These concepts have an implied impact on people, but the connection is loosely defined and leaves many questions unanswered in the minds of anti-frackers. Who will benefit from increased "economic vitality"? Who profits from a strong "energy position"? What do we risk by basing our decisions on abstract values? This contributes to the perception of pro-frackers as lacking heart as the inevitable answers to these questions devalue social equity and environmental protection.

At the same time, anti-frackers exhibit low levels of "Academic Language" and high levels of language that references people. They often motivate vicarious proximity with *pathos*, emphasizing impact on the innocent, be they people or natural places at risk. Due to this, anti-frackers (and environmentalists in general) are often criticized for valuing the sentimental over the practical. Analysis of the vicarious proximity appeals of anti-frackers shows this to be a particularly common tactic to motivate support. One clear example of this comes from Kennedy. The full text of her vicarious proximity appeal is as follows:

> *Children* in Fort Worth, Texas vomit before *they* go to bed at night. There's helicopters flying overhead, there's so much noise, so much traffic and so much water pollution that *these children* are going to bed sick. So *people* who live in *these communities* are afraid, this is not a fear that goes away tomorrow. *Mothers* are fearful of what they are giving to their *children*, the water that *they*'re bathing their *children* in. So you have increased stress on all matters in the community [italics added].

Kennedy populates her vicarious proximity appeal with innocent people who provide the context for her own abstraction. Like pro-frackers, she explains her motivation through use of an abstract concept. In her case, this is the broadly defined notion of "stress" on "all matters in the community." Even Kennedy's choice to describe the impact of these sick and afraid innocent people as "stress on the community" highlights the impact in individual human terms as opposed to broader concepts such as "economic vitality."

These discourse differences between pro- and anti-frackers are in part responsible for the stereotypes each of these groups face. However, a finer-grained analysis reveals alternate uses of vicarious proximity that work to repair these very stereotypes. This surprising finding shows the versatility of place-based appeals and the value of this methodological approach.

Vicarious Proximity as Image Repair

In rhetorical terms, the strength of pro-frackers is in their appeals to *logos* while that of anti-frackers is in their appeals to *pathos*. Though vicarious proximity is at times used in ways that cohere with stereotypes about each group, the major finding of this chapter is that a statistically significant proportion of the time (based on a p value of 0.004 from multivariate analysis), these appeals are used to counter the very same stereotypes. Computer-aided rhetorical analysis in conjunction with close reading of these speeches has shown that vicarious proximity appeals are often used by pro-frackers to be more personal, subjective and to appeal to *pathos*. At the same time, anti-frackers use vicarious proximity to advance *logos*-based arguments that take personal experiences and tend to represent them as objective reality, quite the opposite of what is expected from these groups. To explain how these effects are created, I subdivide my findings according to the following rhetorical features:

Removing versus positioning self into the scene. Anti- and pro-frackers signal different levels of involvement in scenes to create a sense of objectivity (anti-frackers) or subjectivity (pro-frackers).

Science-like discourse as a strategy of image repair is used predominantly by anti-frackers to add validity to their representations of place.

Risk mitigation strategies in the form of personally focused narratives that represent the safety of hydraulic fracturing, used predominantly by pro-frackers.

By studying the differences in the use of vicarious proximity appeals by pro- and anti-frackers, this chapter provides insight into how representing place can be a rhetorical strategy of image repair. The most prevalent way that hearing speakers attempts image repair is through removing the self or injecting the self into a representation of place. DocuScope combined with multivariate analysis reliably differentiates pro- and anti-frackers, but the factors that separate these two groups are certainly surprising. Factors that emphasize *objectivity* are prevalent for anti-frackers (negative numbers in table 4.2), while *subjectivity* is used by pro-frackers (positive numbers in table 4.2).

Table 4.2 shows the results of multivariate factor analysis. While texts show high levels of *First Person* and *Privy* DocuScope categories, they are significantly low (*p* value of 0.004) in *Descriptive* categories. This difference maps onto the texts of pro- and anti-frackers to show that pro-frackers use *First Person* and *Privy* language while anti-frackers tend to use more descriptive language. The categories of *First Person* and *Privy* are both activated when a writer or speaker injects their own personal perspective into a text. The *Privy* DocuScope cluster indicates a passage of text that is one's own

Table 4.2 Results of Factor Analysis Derived from DocuScope Categorization of Texts Showing Significant Difference between Pro- and Anti-Frackers

Group	*Variable*	*Factor (p value = 0.004)*
Pro-Frackers	First Person Privy: Disclosure	0.644
	Privy: Uncertainty	0.493
	Privy: Thinking	0.478
	First Person Privy: Autobiographical	0.451
Anti-Frackers	Description: Sense Property	-0.559
	Description: Motions	-0.479
	Description: Sense Object	-0.476

"private, personal or subjective observation" (*DocuScope*). Quite the opposite, the *Descriptive* cluster is activated by language that describes concrete objects and spaces in an empirically verifiable way. High use of *Descriptive* language is not in and of itself indicative of an objective observer subject; however in this case high levels of *Descriptive* language corresponds with low levels of *First Person* and *Privy*. Still, this finding alone does not fully confirm the argument that anti-frackers use vicarious proximity appeals to illustrate the objective and scientific while pro-frackers take an opposite approach to convey more personal, sentimental accounts of place. Although these DocuScope findings suggest this interpretation, to fully support that argument I combine this analytical approach with the discourse analysis used in previous chapters to closely read each text.

Removing versus Positioning Self into a Scene

To compensate for the stereotype that they lack scientific evidence, anti-frackers tend to remove themselves and other people from their representations of place to augment the appearance of an objective account. This removal of self and other people is often paired with definitional claims about the place. These moves help anti-frackers build their *ethos* as scientific observers. On the other hand, pro-frackers do just the opposite, highlighting their own and other people's involvement in place. As Risa Applegarth explains in her article about the nature writer Mary Austin, asserting the self in writing about place, which she refers to as "attestation," establishes *ethos* and characterizes the writer as a careful observer (Applegarth, 2011, pp. 55–56). Though she writes specifically about the genre of naturalist writing, contrary to Applegarth's claim, I do not see attestation used to assure audiences of the status of the writer as a scientific observer. Rather, in this context it conveys the privileging of people over place. Though *ethos* is always an issue for the general public's entrance into a public hearing, pro-frackers need to mitigate the claims of anti-frackers that they just *do not care*:

> (1) ANTI: "In addition to having a commercial air terminal, there are military units, the U.S. Airways Operation Center, cargo freight business and 5,000-acre free trade zone, a massive fuel farm, five de-icing pads, maintenance centers, an international food cargo hub, a hotel, a fire school and the global headquarters with its own hangar of Dick's Sporting goods. It's already an industrial zone." (Eichenlaub)

> (2) PRO: "I see it every day. You know, we've worked to develop schools, churches, office complex, golf courses, housing development, malls, highways in Pennsylvania. And I can tell you that these operators in this area and the areas

that I've worked in are—talking about it from my personal experience, all do it safely, safely, and environmentally conscious." (Vaccarello)

The two excerpts above serve to illustrate this contrast between pro- and anti-frackers. The first speaker focuses on the buildings and enterprises that populate the Pittsburgh Airport. This extensive list serves to not only allow the audience to create a vision of a crowded airport packed with different enterprises; it also supports the speaker's definitional claim that the airport is "already an industrial zone" (Eichenlaub). Anti-frackers typically make definitional claims in vicarious proximity appeals. Their own experience with the place is downplayed (at the risk of *ethos*) to make their account seem more factual. In this way, anti-frackers attempt to argue with objectivity and with less emotional attachment. In contrast, the second excerpted speaker, Vaccarello, also lists buildings and enterprises for the audience, but provides the list for the purpose of commenting on personal relationships. Vaccarello begins by attesting to his personal involvement with these places in the community and then uses this experience to support his claim about the safety of gas exploration. He emphasizes his own involvement through the personal pronoun, "I," as well as specifically stating that it is his experience (talking about it from my personal experience). In direct contrast to Eichenlaub's similar list of sites within a place, Vaccarello uses this list to encourage his audience to consider the connection between positive community sites and the people who provide those structures. This typical move by pro-frackers to inject themselves and other people into the scenes they depict is used to create a sense of personal attachment to a common place, as we will see in the following examples.

(3) ANTI: "The Upper Delaware River is a wild and scenic river with a inextricably connected watershed blessed with vast expanses of unbroken forest, agriculture, rural landscapes, trout streams, historic settlements and towns." (Carluccio2)

(4) PRO: "To mention South Park, when I grew up, it was my playground. I learned how to golf very badly there and swim, ice skate. And it has really disheartened me to see the decline in conditions in not only South Park but all of the nine county parks. This is through no fault of the county or of its staff. It's a condition based on poor economics." (Coyne)

With few exceptions, anti- and pro-frackers ascribe homo-centric value to the places they discuss. That is, the value of a place is in its use as a resource to humans (Killingsworth & Palmer, 1992). Though the anti-fracker (3) Carluccio argues for the natural grandeur of the Upper Delaware River, he relates the importance of this place in human terms. Agriculture is a human process; landscape, a human vision of a natural place; trout streams, a source of food and recreation; and historic settlements, sites to understand ourselves as humans. These examples all create a natural, distinctly human scene. Likewise, pro-fracker Coyne argues for the importance of South Park based on its use for recreation and explains its dilapidated state via economics. Whether they are describing a natural place or a place that has been specifically created for human use, the value of place in public hearings is almost always positioned in homo-centric terms.

Where these two sides differ is in the degree to which speakers inject themselves into these appeals. Excerpt (3, anti-) and (4, pro-) show patterns similar to what I established above, that anti-frackers remove and pro-frackers insert themselves into their depictions of place. In (3), the speaker creates a sweeping depiction of the Upper Delaware River full of objects of human importance but barren of humans and their subjective impressions. (3) Carluccio's evaluation of the positive aesthetic value is apparent through his choice of the terms "scenic river," "blessed," and the negated negative "unbroken forest" to emphasize the rarity of such a thing. He bears witness to a place his audience should consider objectively rare and valuable and uses synecdoche to convey it, conforming to the patterns we saw in chapter two. Despite this evaluation, Carluccio himself is nowhere to be found in this depiction and is only an implied first-hand observer, which is typical of anti-frackers.

In contrast to (3) Carluccio, (4, pro-) Coyne places himself in direct relationship to the place he describes. Not only does he include himself by use of the first-person pronoun, but he also lays claim to and names aspects of the Park specific to his use (my playground). In addition, Coyne explicitly states that he is "disheartened" about developments in the Park. Rather than support fracking based on the variety of other lines of argument available to pro-frackers, he chooses to use his three minutes to establish his personal connection to South Park. This focus on personal connection runs counter to stereotypes of pro-frackers that suggest they support gas extraction based purely on its economic merits:

> (5) PRO: "I live in Pittsburgh. I'm a retired firefighter. And I don't have any statistics. I don't have any fears. I just have a little bit of common sense. I grew up in Pittsburgh. When I was young, the community streetlights were on. Every morning when you'd wake up, there'd be a half an inch of chemical dust on your car. We got over that. I had asthma as a child. I don't have asthma anymore. The air in Pittsburgh is good. The water is good. I can bass fish at the Point. I couldn't do that 40 years ago. The laws that we have in this state protect the environment, and they're being enforced. They will take care of this industry,

> but we need the energy now. So as far as I'm concerned, it's drill, baby, drill." (Clark)
>
> (6) ANTI: "How many of you have actually seen an operational well pad? Okay. And how many of you have talked to the people in the Woodlands? They—their wells have been polluted by drilling. They have no clean drinking water. So that's an example of drilling coming very, very close to where they live and your wells have gone bad. I'm definitely not a genius, but I do have common sense, and common sense tells me that it's a very bad idea to build well pads near an airport." (Supowitz)

Even in cases where anti-frackers use vicarious proximity to focus on negative impacts to people, they tend to be absent as characters in the scenes they create. Alternatively, pro-frackers consistently create appeals that emphasize their own involvement. Comparing excerpts (5, pro-) and (6, anti-), this trend is clearly seen; Clark in (5) supports fracking by appealing to common sense based on the parallel images of having asthma in a polluted environment as a child and having recovered as an adult. Clark goes much further in creating an image of historic Pittsburgh than (6) Supowitz, who briefly signals polluted wells and drinking water before calling attention to the proximity of well sites to residential spaces.

Although Supowitz grounds her claim in common sense using a similar construction to Clark, the degree to which each creates an objective observer's perspective differs. Throughout Clark's speech, he focuses on his own subjective experience. Foregrounding himself, he indicates that his experience is representative of changes seen by others and in other places. On the other hand, Supowitz addresses her audience and leaves her own experience with well pads and with the people affected by them implied by asking about the Woodlands. She also invokes people from the Woodlands, whom she strongly suggests have been negatively affected by fracking. She creates the experience of an arbiter of fact about the cause of polluted wells and the danger of close proximity between well sites and houses, while Clark is much more concerned with developing the first-hand nature of his account.

The accounts of places of pro- and anti-frackers do not simply differ in the amount that they focus on people. The presence or absence of self in vicarious proximity appeals creates a different experience of the speaker. Pro-frackers explicitly thematize their own involvement in the scenes they create for their audience. Anti-frackers may populate their vicarious proximity appeals with people, but they tend not to include themselves in these depictions. Vicarious proximity is used to create a sense of personal investment for pro-frackers and the *ethos* of an objective observer for anti-frackers. For both, vicarious proximity appeals provide an opportunity for speakers to connect with their audience through a guided visualization of a place. Though vicarious proximity is created through features common to both pro- and anti-frackers, their emphasis on the first-hand nature of their account results in the surprising difference that pro-frackers are more subjective and personal in their accounts of place while anti-frackers are objective and removed from the scenes.

Science-Like Discourse for Image Repair

To repair their image and further establish their ethos as an objective observer, anti-frackers borrow from scientific discourse. As industry-funded studies on fracking are amplified by pro-frackers, anti-frackers scramble to oppose research that they believe is flawed. This is especially the case in the earlier hearings, when academic research on this topic was not as common as it is today.

As I have shown above, by removing themselves from vicarious proximity appeals, anti-frackers attempt to create a sense of objectivity to the scenes they invite their audience to experience. This sense of objectivity, achieved by removing the self from vicarious proximity appeals, is enhanced by the use of grammatical structures similar to those common in scientific discourse (Halliday & Martin, 2003). Scientific discourse is privileged in these public hearings because people turn to scientists to better understand the risks of hydraulic fracturing. By speaking like a scientist, anti-frackers can be said to use *bolstering*, a strategy of apologia through which "a speaker attempts to identify himself with something viewed favorably by the audience" (Ware & Linkugel, 1973, p. 277). Speakers attempt to create a likeness between their speech and that of a scientist. If successful, they have a better chance of communicating their perception of the risk posed to their communities.

Anti-frackers may borrow this scientific discourse to ascribe the status of scientific knowledge and objectivity to their words. Halliday and Martin (2003) explain that "the writer is following a fashion by which he seeks (unconsciously, in all likelihood) to give extra value to his discourse by marking it off as the discourse of an intellectual elite" (p. 70). Though speakers in public hearings may not go so far as to claim that they belong to an intellectual elite, several markers of scientific discourse identified by Halliday and Martin do increase credibility and the likelihood of veracity in the claims made. Halliday and Martin explain that scientific discourse is marked by high levels of *lexical density*, a measure of the number of content words in a sentence (e.g., "glass crack growth rate"; 2003, p. 76). This, in part, contributes to the larger category of the *grammatical metaphor,* which is characterized by the nominalization of processes (p. 79). Grammatical metaphors are essentially substitutions of one grammatical class for another, in this case verbs for nouns in nominalization. Nominalization helps scientists to explain step-by-step processes so that what is already established and what follows are clearly presented as separate clauses (e.g., "Each *absorption* marks its transition to a state of higher energy, and each *emission* marks its transition to a state of lower energy"; p. 81). The use of lexically dense phrases and the nominalization of processes to construct grammatical metaphors are recognized rhetorical hallmarks of scientific discourse. The same lexical density and preference for nominalization can be found in anti-fracker public hearing discourse as speakers argue for the credibility of their claims.

Out of the speakers who oppose fracking, 61 percent make use of lexical density within vicarious proximity appeals to argue for the veracity of their perspective. In anti-fracker texts where lexical density occurs, it appears at a rate of 1.74 instances per text. This is opposed to pro-frackers, who use lexical density at a rate of 1.43 instances per speech out of the 46 percent of speeches where it occurs. The difference between anti- and pro-frackers' deployment of lexical density is not only a matter of rates. Anti-frackers include lexically dense (defined here as a minimum of two modifiers of one noun, excluding proper nouns) to establish scientific fact while pro-frackers tend to use lexically dense phrases to reference industrial and commercial places (e.g., "large gas field area" and "bustling little communities") or financial instruments (e.g., "oil lease bonus monies" and "windfall profits tax").

When an anti-fracker says that the Upper Delaware River is an "inextricably connected watershed" (Carluccio2) or that "Cohocton is one of the highest water quality rivers" (Trondsen), we see high levels of lexical density at work to establish the claim as fact by including grammatical structures common in scientific discourse. Rather than focusing on how water quality is determined (which the speaker is likely unable to explain due to their lack in expertise), they borrow the lexically dense grammatical metaphor of "high water quality river" to establish fact. Other examples from anti-frackers include "appropriate fracking fluid," "low summer [river] flows," and "[arsenic] 6,430 times acceptable level." Though Halliday and Martin's interest in scientific discourse is rooted in non-native English speakers' experience when reading scientific texts, these same structures seem to become portable units readily recontextualized by nonexperts for public use.

Anti-frackers use a grammatical form that helps establish their claims as scientific fact. Besides the argumentative purpose of all vicarious proximity appeals, which is to bear witness to a perceived change or risk, many nonexperts who attempt to create a sense of scientific reasoning emphasize the complexity of nature. They argue against fracking, assuming that it would disrupt a natural order that humans do not fully understand. When Carluccio calls the Upper Delaware River Watershed "inextricably connected," he implies that a failure in one small area would affect the entire system, relying on synecdoche to advance this argument.

Even when speakers do not use lexical density to their advantage to create a sense of scientific reasoning, they often argue for a view of nature as beyond full comprehension. For instance, Greenberg cites the "very complex nature of nature" and Hyman says that "our water is an intricate part of the Delaware River Watershed." Speakers invoke complexity and intricacy to base claims in scientific reasoning, but they convey a general distrust that we as humans know enough to predict the impact of an intervention like hydraulic fracturing. So, anti-frackers often work to repair their image by rhetorically aligning themselves with scientific reasoning, but they then use this position to argue that current science does not allow us to fully understand something as complex as nature. To interpret nature through the lens of scientific rationality while emphasizing its ultimate incomprehensibility, anti-frackers invoke apologia via *transcendence,* providing a larger context with which audience members are already familiar (Ware & Linkugel, 1973, p. 280). The contrast between the knowability of science and the unknowns of nature helps support a rhetoric of postponement. Apologetic transcendence facilitates the repair of the anti-fracker image by connecting the places they describe to scientific rationality and to the unknowns that remain. This, in part, factors into how anti-frackers communicate risk.

The assessment of acceptable levels of risk is dependent on both scientific and social rationality; the communication of risk in a public hearing is no different. Ulrich Beck explains, "The scientific concern with the risks of industrial development in fact relies on social expectations and value judgments, just as the social discussion and perception of risks depend on scientific arguments" (Beck, 2003, p. 30). What we see in public hearings on fracking supports the latter half of Beck's claim. Speakers in the hearing convey risk in a way that is dependent on science. However, "science" in this setting is not necessarily the rigorous attention to method and replicable results normally associated with that venture. "Science" can be invoked by grammatical construction and attention to phenomena that are often the purview of science. Anti-frackers use vicarious proximity appeals to argue for unacceptable levels of risk that are grounded in what they present as the product of scientific rationality.

Though it is very unlikely that audience members are mistaking the claims by nonexperts for those of scientific experts, nonexpert presentation of evidence that relies on science-like rhetoric at the very least gives credibility to their claims. For instance, Josh Fox, the creator of *Gasland*, spoke of his research at a 2008 hearing in New York State:

(7) ANTI: What I heard was absolutely devastating health effects on populated areas, health effects on forests, health effects on water. I want to talk about human health, human health for anybody within 2,000 feet of a well pad because

> it's probably about eight percent of this room. I saw fracking done that was being sprayed over landscapes, in groundwater, in aquifers and thousands and thousands of trucks. Trucks sworm [*sic*] every 30 seconds, you can stand on any corner—well, there will be a lot of new corners, but you will see a truck go by you every 30 seconds. None of the sites I filmed in Wyoming, Texas or Colorado, I filmed hundreds that didn't have inspectors at all. They were completely overrun. The states cannot keep up without the Federal Acts protecting this or monitoring or regulating as you heard. 16 inspectors in New York State to monitor 20,000 wells, it's impossible to do that. (Fox)

Fox does not rely on lexical density to create the persona of a scientist, but he does invoke scientific rationality in several ways common to anti-frackers. He uses quantitative data to support his claims, even if the numbers he cites may be somewhat suspect. Fox gives specific numbers (eight%, 30 seconds, 16 inspectors) based on what he projects as observable data. Additionally, he does not give a value-based assessment of acceptable levels of risk as Beck's social rationality would lead us to expect. Instead, he presents widespread negative health effects (despite being unable to provide more specific information about those health effects), descriptions of the hazardous treatment of wastewater, and well sites that are "overrun" presumably with workers and absent of inspectors. Though Fox inserts his personal perspective into this account through frequent use of the personal pronoun "I" (uncommon among anti-frackers), he contrasts his observations, which he suggests are factual, against pro-frackers' claims of safety.

Risk communication, in this case, is not only transferred from the top down. Anti-frackers like Fox assume the mantle of scientific rationality and communicate their "findings" by way of invited visualization that often includes rhetoric characteristic of scientific discourse. Others, such as Colucci from the 2014 hearing and Miller from 2008, create a sense of risk by claiming to represent a larger problem. Colucci depicts a well explosion that happened in Avella, Pennsylvania while Miller presents a hypothetical account of the transmission of fracking chemicals into ground water:

> (8) ANTI: In some industries these chemicals are highly regulated and my experience is that any time something is released into the environment these wells, I don't think are closed systems, these chemicals are going into the earth, some of them are staying in the earth and they are going to move, they are not going to stay some place far away from us. I believe they will end up in our groundwater, that is very likely. (Miller)

> (9) ANTI: Avella, PA, on a developed property where Atlas Energy was engaged in hydrofracking of wastewater impoundments where the fracking fluid

> makes its way back to the surface is stored and it exploded flames went 200 feet in the air, burned for six hours and produced thick black smoke, a cloud visible ten miles away. Soil sampled onsite was found to contain arsenic 6,430 times permissible level. Pretty impressive. (Colucci)

Out of all the speakers in these hearings few are actual scientists, though anti-frackers use the discourse of science to add credibility to their evaluations of excessive risk. Contrasting Miller's speech with Colucci's is especially interesting considering that Miller prefaces his comment with the claim that he is a trained chemist. As an actual scientist, Miller discusses a scientific process in excessively pedestrian terms, while Colucci, not a trained scientist, provides specificity and quantitative evidence that is more akin to scientific discourse. Colucci refers to "acceptable levels" of a specific chemical (arsenic) and the collection process used to gauge the presence of arsenic. Miller adjusts in the opposite direction. He argues for the risk of contaminated groundwater by removing all scientific language from his speech, even though he is likely quite capable of describing the process of chemical transmission to groundwater in more precise terms. These two examples show how speakers adjust for their audience relative to their own position. Maligned for lacking scientific evidence, anti-frackers are prone to use the discourse of science to bolster their claims about risk.

Authority and Personal Accounts in Risk Communication

Anti-frackers have been shown to cultivate authority by using science-like discourse when they communicate risk. When pro-frackers participate in discussion of risk, they follow the pattern of image repair established in this chapter by injecting themselves into scenes and providing personal accounts laden with the language of caring and connections between people. Pro-frackers are less descriptive in the scenes they create, foregoing setting for strong characterization. Through this they work to establish authority by presenting themselves as stewards of the environment and caretakers of their communities; they use personal experience to speak not only for themselves, but for their neighbors as well.

Pro-Frackers as Environmental Stewards

One main area of repair for pro-frackers is the perception that they do not care about the environment or, at the very least, that they value economic vitality over the environment. Pro-frackers answer this criticism by directly discussing their affinity for nature and by mitigating environmental risk from fracking. Pro-frackers, in 31 percent of their speeches, foreground concern for the environment or environmental protection measures (e.g., "We can all agree that we are concerned about the environment" or "gas companies that we work for in developing this shale play go above and beyond, above and beyond the environmental protection that they do on these sites"). Dan Bagley, who works in the gas industry, explains his affinity for nature by characterizing himself as an "outdoorsman" and a family man who "demand[s] clean water for [his] children." He extends his personal experience to his co-workers and speaks authoritatively on their behalf to mitigate the risk of environmental degradation. He bases this on their understanding of the process and their familiarity with those who are involved.

> (10) PRO: I'm an outdoorsman. I cherish the waters and the lands of this state. Our families are avid fishermen, as you can tell from the camouflage CRI caps that we wear. I was raised in Arkansas amid oil wells and gas wells in the natural state, where energy revenues and land management were a boon to wildlife and waterway cleanliness. We demand clean water for our children, too. Many of us are well-water-fed. We do not fear natural gas development because we understand it, and we know the people who perform the work, some of whom are with us at this meeting. So safety is important to us, too. (Bagley)

By citing personal experience, Bagley persuades the audience to trust him. He does not attempt to explain natural gas development so that his audience can understand and not fear it. Instead, he creates an emotionally laden appeal that asks his audience to trust the industry because they "cherish" the natural environment and because their children drink the water. Analogous to this situation, he presents his experience in Arkansas, explaining that development, though counterintuitive, can have a positive effect on the environment. His focus on interpersonal relationships creates a strong sense of pathos, which is usually associated with the other side. Leaving science out of the equation, he foregrounds trust in individuals who care about the environment. This strategy not only works to mitigate risk, but also repairs the pro-fracker image of being overly economical and little concerned with the environment.

Personal Experience and Heritage

As explained, an emphasis on personal experience with hydraulic fracturing, along with family and heritage, helps pro-frackers mitigate risk. They insert themselves as subjects in 84 percent of vicarious proximity appeals (i.e., use of "I," "me," "we," etc.). They refer to heritage by explicitly situating future generations or family history in a place in 34 percent of speeches. This combination helps repair the image that pro-frackers are out to make a quick buck without considering the future of where they live. This repair is often accomplished by creating vicarious proximity appeals that center on rural places, such as farms. Mike Yillit, an employee of the fracking industry with family ties to agriculture, presents one such case:

> (10) PRO: My days start at 5:00 in the morning. I also own a farm that is our family farm, along with some of my cousins and uncles. We also have our farm ground leased for Marcellus Shale. The well pad is about 400 feet from our family home. It belongs to my sister now. It will be handed down to my kids. There will always be a farm. There's about 20 farms in this area that we live that all have been drilled and fracked. The fracking process on one well takes about two days. I handle the chemicals. Everybody talks about—there's no regulations. The chemicals that I handle, if there's an incident on the road, I can be in prison. If you're on the Turnpike and a semi upsets, that water goes—that fuel goes into the storm drain, that's a hazard. Every time you folks use salt on these roads, you create brine water. We have 20 farms in this area. Our water comes from two wells and a spring. (Yillit)

Yillit provides a light description of his farm and the well sites located on his property (including the proximity of the well to their home). Instead of a strong description, he follows the pattern established by other pro-frackers to emphasize their personal connection to the land. He accomplishes this by adding characters to his scene: his cousins, uncles, sister, and children. The children are used to create a heritage appeal in which he states, "[The farm] will be handed down to my kids," and he points to the durability of this scene by saying, "There will always be a farm." The emphasis on family heritage creates a sense of certainty about the safety of hydraulic fracturing. This is combined with Yillit's claims about his work in the industry and the severe penalties he would face if he did not follow regulations.

Yillit not only mitigates risk by citing the durability of his family farm and his concern for the people who live there, he also extends his claim to those in the surrounding area. Though Yillit's speech is somewhat disjointed, he clearly claims that twenty farms in his area have been fracked and that all of their water "comes from two wells and a spring" that he implies provide safe drinking water. He does not say whether or not that water has been tested but implies that any risk faced from fracking does not affect his family or the others around him. By citing other instances of water degradation, that is, brine water from road salt, Yillit acknowledges the possibility of contamination, but maintains his confidence in the fracking process.

As pro-frackers create alternate representations of their lives as environmentally minded and invested in the future of their community, they participate in the *justificative* posture of apologetic rhetoric (Ware & Linkugel, 1973, p. 283). Pro-frackers not only wish to create a sense of understanding around fracking, they also seek their audience's approval of natural gas exploration. Claims to pro-frackers' interest in the environment and heritage are, in part, designed to transcend what may otherwise make fracking seem offensive (Benoit, 1997, p. 181).

IMAGE REPAIR AND APOLOGIA USING A BLENDED ANALYTICAL APPROACH

In this chapter, I have focused on the discourse in multiple public hearings in order to understand how the use of vicarious proximity differs between those for and against hydraulic fracturing. Based on the findings of the previous chapter that show vicarious proximity to be a ubiquitous rhetorical resource for participants in these hearings, I looked at how the critical features of these appeals could be differently emphasized to create various effects. The ensuing comparative analysis of vicarious proximity in shale gas hearings supports several significant claims about the use of vicarious proximity, its value as a unit of analysis, the value of computer-aided corpus analysis for this type of investigation, and the role of vicarious proximity in ordinary democracy.

The major finding of this chapter is that vicarious proximity is used as a strategy of image repair by both sides of the debate. Confronted with stereotypical representations of their side, both anti- and pro-frackers respond to these attacks by using vicarious proximity appeals to advance an image of themselves and their relationship to the land in question that runs counter to stereotypical representations of their side. What is most notable about image repair in these hearings is that it is not done as an explicit response to an accuser, as is standard in typical cases of image repair and apologia.

My work here extends the genre of apologia to include cases previously thought outside of the realm of apologetic discourse due to their lack of direct accusation and explicit strategies of image repair. What I show is that apologetic discourse can be created by implicature and lexico-grammatical choices. This analysis also provides further support for Downey's (1993) identification of "dual motive apologia" in which rhetors not only repair, but also enhance their own image. In this sense, apologia functions not just as defense, but also as offense. In the setting of the public hearing, this makes sense. The genre of this rhetoric, first and foremost, is deliberative, which results in apologia that is largely implied.

One reason that image repair is implicit rather than explicit in these public hearings has to do with the contextual constraints of the hearing itself. Speakers address the council members and other politicians seated at the front of the room. They are not to address the audience, so rejoinders with the crowd may detract from their message. However, through vicarious proximity appeals, which ask the audience to imagine a scene and accept the judgment the speaker passes on that scene; speakers make choices that nevertheless address accusations while allowing them to maintain focus on the front of the room. These choices are not overt strategies of the type that Ware and Linkugel or Benoit associate with apologia and image repair, which result in complex arguments, created in long spans of discourse. Instead, speakers make choices regarding whether or not to appear as subjects in the scenes they depict, the lexical density of their speech, and the implied relationships between people and places to address the concerns of those who would accuse them of being misguided. More succinctly, vicarious proximity appeals provide a vehicle for complex representations of group identity via representations of place. The relationship between place and identity has been well discussed (Bamford et al., 2014; Clark, 2004; Farbotko, 2005; Johnstone et al., 2006; Larson & Pearson, 2012; Modan, 2000), but this view of place identification as a strategy of image repair has not. Most interestingly, image repair is not the primary argumentative purpose of these appeals. As I have shown in the previous chapter, vicarious proximity is used to establish ethos, communicate risk and evaluate transformations of place. However, image repair functions as a somewhat covert strategy. The narrow focus on vicarious proximity appeals has revealed an important strategy that otherwise might have been overlooked due to the more explicit ends of these appeals.

Democratic Participation, Image Repair, and Common Ground

The public hearings I have studied operate under formalized patterns of interaction of speaker-to-council with very little two-way communication and prohibited communication between speaker and audience. This differs notably from the school board meetings full of direct argument that helped Karen Tracy develop her concepts of "ordinary democracy" and the related "reasonable hostility." Nevertheless, both public hearings and school board meetings are sites of ordinary democracy. This study of public hearing discourse provides a different view of reasonable hostility that shows how citizens can address their own or others' hostility while operating under different communicative constraints. Central to the idea of reasonable hostility is that when people are angry yet communicate in a respectful way, democracy is working (Tracy, 2010, p. 203). Reasonable hostility facilitates "the impulse to build an ideal that recognizes competing values" (p. 203). Because participants in these hearings do not belong to a "strong public" (Fraser, 1990), but have no direct decision-making power, they do not actively work to "build an ideal." However, in part through the image repair strategies highlighted in this chapter, these hearings reveal competing values.

Whether or not the sides of this debate would agree that they hold similar values, their discourse, when viewed through the lens of vicarious proximity, shows that each side works to repair their image by emphasizing values that the other purportedly places in high regard and that they do not. Through implicature and narratives that present counterexamples to stereotypes, participants complicate the value hierarchies typically ascribed to them. In other words, through strategies of apologia like bolstering and transcendence, speakers proclaim that they actually hold values similar to their opposition. Despite these same values, they have reached a different conclusion on the issue at hand. Therein lies the rub, as it seems unlikely, even if both sides recognized that they were arguing from the same set of values, that they would persuade one another to change sides. Still, that is not the point of the public hearing. Ideologically ensconced citizens attempt to persuade a council that is, at least formally, undecided about a course of action.

Through strategies of image repair, both implied and explicit, citizens come closer to establishing common ground despite their fundamental disagreement. Both sides value scientific evidence yet are also personally invested, both sides worry about safety in relation to their neighbors and their families, both sides are concerned for the future, and neither side wants to see people suffer health or financial problems. Nonetheless, misunderstandings and unfair stereotyping creates the impression that one side or the other does not care about these issues. The public hearing format is not ideal for allowing a healthy degree of reasonable hostility. Other formats for these public meetings, such as one based on principles of deliberative democracy (Barber, 1984), would allow for more direct challenge to ideological positions, though it is not clear that citizens would attempt image repair to the same degree. It may be that the planned speeches that are typical in these hearings allow participants to present narratives that run counter to accusations and stereotypes in a way that remains reasonable and not overly hostile.

As I have shown, despite the lack of direct engagement, speakers do make sophisticated rhetorical appeals both in a deliberative sense and as image repair strategies that are targeted toward their audience. Anti-frackers shape their discourse to mimic scientific language, presenting themselves as arbiters of fact. Pro-frackers highlight their investment in the places their families and communities inhabit. We can see these decisions on representing place as audience-focused moves that serve multiple purposes. Anti-frackers, for instance, advance deliberative argument about why fracking should be stopped through their observation of a drilling site. In the very same appeal, they make an identification move to help like-minded audience members embolden their opposition as well as respond to criticism of their position as not scientific through objective representations and by drawing on scientific discourse.

The limitations of public hearings may prevent them from becoming ideal sites of democratic participation. Although the speeches that participants give have an impact on the democratic process of the audience, the audience does not necessarily engage with the different perspectives that confront them. Despite limitations, we have seen that citizens contribute to publics in meaningful ways that advance their concerns, employ identification strategies based in common places, and address the criticisms of those who oppose them. Uncovering the rhetorical strategies common to both sides allows us to see how a mixture of common experience, ideology, and circulating discourse results in shared strategies.

Vicarious Proximity as a Unit of Analysis

It is clear that at the heart of the disagreement between these two groups, no matter the variety of their arguments for or against (based in science, history, common sense, and personal experience), both sides argue to define the identity of the sites under consideration for hydraulic fracturing and the communities attached to those sites. Modan (2002) explains how "social actors actively construct fluid identities for places through discourse" (p. 489). This is often accomplished through variations in language structure (p. 488). My focus on vicarious proximity appeals has allowed me to examine linguistic variations within a single topos of public discourse, the vicarious proximity appeal. Had I considered the full speeches of each participant, the number of variables associated with differing argumentative purposes, contexts, and language structure would have obscured the significant differences found between anti- and pro-fracker discourses.

Though vicarious proximity appeals have proved a useful unit of analysis, questions remain as to how far these results can be generalized. I expect that further analysis with a broader scope would show similar strategies of image repair in other types of appeals used by both sides. I also expect that these results would apply to other environmental disputes and perhaps beyond.

As a method of analysis, computer-aided corpus rhetorical analysis has provided insight that would not have been possible using other methods, but there are shortcomings to this approach as well. The advantages are that minute differences between corpora, using a program like DocuScope in conjunction with statistical analysis software, can be brought into relief with relative ease. In my case, the significant difference in objective and personal language use between anti- and pro-frackers caught me by surprise and may have been overlooked were it not for the digital tools. This method, like others, does not provide an interpretation of findings. In this sense, computer-aided analysis provides something of a (strong) hunch that must be substantiated by other methods. Following computer-aided analysis with close reading aided by concepts from rhetorical and discourse analysis allowed me to consider the purpose of statistically significant micro-differences between corpora. The shortcomings of this approach are the difficulties it presents when working with public hearings, both in terms of access to transcripts that can be formatted as well as the length of speeches. DocuScope works best with texts that are longer than 500 words, but many of the vicarious proximity appeals that I analyzed are far shorter than that. To counteract imbalances in findings, it was necessary to combine all of the speeches of one side into a single file in addition to the standard procedure of analyzing a large corpus of individual speeches. This obscures differences between speeches but allowed me to confirm my findings from the analysis of individual speeches.

CONCLUSION

In this chapter, I have shown that place-based appeals contain multitudes, from explicit appeals for or against a change to the land, indicators of identity and group affiliation, to micro-rhetorical resistance to stereotypes. Public hearing speakers tap into the deeply important role of spatial memory in the human brain when they vividly recount experiences in place. In surprising ways, rhetors shape these representations to strategically advance multiple rhetorical aims. Those who agree on issues tend to represent place similarly. This suggests that depending on where you stand, you really do see a different world. Shaping how people see the same part of the world is the way to win them to your side.

Chapter 5

Conclusion

Themes, Challenges, and Findings for Public Place-Based Rhetoric

I close the final chapter of this book with two simple claims about Marcellus Shale gas extraction that are indisputable: (1) Hydraulic fracturing impacts the land where it occurs and (2) the people who live in varying degrees of proximity to these industrial sites experience this impact. The dispute begins when we question the extent to which people and place are affected as well as how much people are willing to risk for financial reward. Some believe the risk of environmental degradation to water and air quality is negligible compared to the benefits gained by extraction. Others view the potential benefits of fracking as not worth the risk. On the surface, this simplifies a complex issue, but by starting with the most obvious common ground and examining the discourse used to step beyond that zone of agreement, we better understand the range of values, commitments, concerns, and places that make up our rhetorical terra firma.

In public hearings focused on this issue, participants range from unbridled enthusiasts for a revitalizing force in stricken communities to doomsayers that warn about devastation to their health and shared places. Both sides claim that the repercussions will be felt for generations to come. No matter the side, at the center of many of these public appeals is a deep concern for place. For good reason, we can accept the claim: "Every language is located in a space. Every discourse says something about a space (place or sets of places); and every discourse is emitted from a space" (LeFebvre, 1991, p. 132). To LeFebvre, we use discourse not only to comment on place, but also to construct it. The construction of place is, in many ways, the primary work of hearing participants as they implicitly answer the question: *How should we value our land?* Breaking this into its constituent parts can help us see the points of contention and the issues at stake to simply provide an answer.

In the question, *"How should we,"* the use of the modal verb "should" shows the belief that humans have dominion over the land because we assume the right of intervention. Nonetheless, the wording also recognizes that we create nature. It is our decision to determine how to alter or preserve the places that, in part, provide the backdrop of our lives. Place is not a mere setting. Neither is it objective reality that we experience the same way. Our experience of place, especially the natural world, "reflect[s] human judgments, human values, human choices" (Cronon, 1995, p. 34). Though the word nature itself implies something unshaped by humans, participants in the hearings I have studied do not see the natural world as an agent. Speakers overwhelmingly frame this issue as one of human control of nature. Impact is likewise framed as impact on humans much more than it is framed in ecological terms.

This anthropocentric view of humans' relationship to nature is not surprising, but investigating the type of relationship to nature that hearing participants represent can uncover the ideological foundations of disagreement. How might we imagine this relationship to nature? Killingsworth and Palmer (1992) offer a continuum ranging from "Nature as Object" to "Nature as Resource" to "Nature as Spirit" (p. 11). Within these categories, the authors place enterprises along a continuum from traditional or mainstream science, aligned with Nature as Object, and *deep ecology* at the extreme opposite, with Nature as Spirit. A deep ecology perspective is the only one on this continuum that views nature as both an agent and as important as humans. All other perspectives including science, government, business, agriculture, and social ecology (humanistic environmentalism) are anthropocentric.

Though most people view nature in terms of what it can provide for us, Killingsworth and Palmer's continuum shows we can interpret the world differently even while maintaining our dominant perspective. We do not have to lock into one way of seeing nature. Despite this range of available relationships, we tend to flatten these distinctions (p. 12). Nature is a resource in the form of natural gas or hiking trails. Both imply dominion, though in varying degrees. In public hearings, speakers reveal their dominant attitude toward nature while they categorize and evaluate others. Cronon (1995) explains that people "pour into that word [nature] all their most personal and culturally specific values: the essence of who they think they are, how and where they should live, what they believe to be good and beautiful, why people should act in certain ways. . . . The result is a human world in which these many human visions of nature are always jostling against each other." (p. 51). Despite these competing values, no one argues against the importance of a healthy natural world and, as shown in chapter four, many pro-frackers emphasize their relationship to nature. Accepting that our perspective on nature's value is filtered through anthropocentrism, we may still hope to find contact points across human ideological divides to a more ecological perspective.

The next part of the question, the *value* of land, can be considered in terms of the monetary value that energy companies are willing to pay for a lease, the recreational value of park space, the inherent value of the wilderness, the value of protecting land so future generations can experience it as we have, and many other ways of attributing value to place. Focus on values also shows how speakers appeal to different audiences by framing their relationships to places and each other. Perelman and Olbrechts-Tyteca (1969) explain that values are the "objects of agreement in regard to which only the adherence of particular groups is claimed" (p. 75). This means that values would be indistinguishable from truths if not for the particularity of how they are defined for specific audiences. The universal audience may be targeted by values assumed to be held by all, such as "the *Good*," but Perelman and Olbrechts-Tyteca explain that such values may appeal to a universal audience only when they are not specified. They clarify, "As soon as we try to go into details, we meet only the adherence of particular audiences" (p. 76). In Marcellus Shale public hearings, we may view all speakers as arguing for the common good, but attention to particular values and value hierarchies offers a view into what drives groups who are ideologically aligned and what is common across factions. This may also be viewed from a public perspective to uncover factions within the public convened in a hearing.

The last part of the question, *our land*, prompts us to reconsider who can make decisions about what is done with land. Issues like the spotted owl controversy in the Pacific Northwest and the controversy over defining wetlands show that even land that is privately owned may still be conceived as "ours" (Moore, 1993; Schiappa, 2003) At the same time, land that is legally "ours," like public parks and state forests, can be leased to private interests for industrial activity that transforms our understanding of it from a place of recreation to one of industry.

At its essence, we engage with the shifting definitions of what is public and what is private. Private land owned by an individual, when it is connected to public land by streams, can severely impact what everyone owns. Defining "our land" also allows us to question the jurisdiction of public engagement that takes place at a county level but invokes national and international dilemmas. Where is our action best located? *How should we value our land?* We answer this question through our actions, relationships, and physical imprints on the surface of the earth, but we grow it into a conversation through our rhetoric.

I am interested in society's use of signs that represent and construct space more than I am in the actual spaces themselves. Representations of space are a system that can be deciphered (LeFebvre, 1991, p. 33). LeFebvre distinguishes representations of space from spatial practice and representational spaces as "a system of verbal (and therefore intellectually worked out) signs" (p. 39). This focus has enabled insights about the role of place in public argument and the rhetoric of place as it is used to support or oppose this issue. I have found significant differences between how pro- and anti-frackers use place as an argumentative resource. These findings do not hold for each and every pro- or anti-fracker, but do represent significant tendencies for each group. By uncovering these tendencies, we can understand common rhetorical strategies used in deliberative argument of environmental disputes: strategies of ideological defense (*apologia*), the ways groups attitudinally align themselves and others in relationship to nature, as well as how speakers communicate risk and construct ethos.

TENDENCIES OF CITIZENS IN ENVIRONMENTAL CONTROVERSIES

The analogy used in chapter 3 to describe the difference between pro- and anti-frackers' representations of place can be extended across the findings in these chapters. Pro-frackers provide an aerial view of place, while anti-frackers provide an on-the-ground view. Pro-frackers are less prone to depict a single place, whether it be a drilling site, a natural place, or a home. To separate places as "single" is a difficult venture, but what I intend for this idea to capture is that pro-frackers tend not to represent a place that can be apprehended by a human being standing on the ground. Instead, pro-frackers refer to multiple drilling sites at once, large swaths of forest that cover hundreds of acres, and geopolitical places as home (such as the state of Pennsylvania instead of an individual family dwelling). They take a big picture, zoomed out view.

By representing place in this way, pro-frackers support arguments that nature is resilient enough to be minimally disturbed by fracking. They suggest that the sheer number of fracking sites prove that they are safe. By making claims about large-scale place, they attempt to command authority over those places. I have shown in chapter two that this type of representation corresponds with a whole-to-part use of synecdoche. That is, large-scale representations of place subsume the small and command authority. When fracking worker Mike Yilit claims, "I've worked all through Pennsylvania. I do high-pressure fracking. I've sat on the iron when they fire the gun to frack pits. You don't know these things," he utilizes this broad geographic perspective not only to build ethos, but also to advance superior knowledge. Although Yilit does bring a close view to the process of fracking, using industry jargon ("the iron," "fire the gun," "frack pits"), he keeps his representation generic to any particular experience.

This insertion of the self into scenes to provide first-hand accounts is characteristic of pro-fracker discourse. At much higher rates than anti-frackers, pro-frackers represent scenes from their own perspective. I argue in chapter 4 that this is done to repair the image that pro-frackers are only financially motivated and do not care about people. Through their place-based rhetoric, pro-frackers present themselves as connected to the natural world and concerned with its preservation, but convinced by its resilience so as to not hamper resource development.

Though I have limited my study to hydraulic fracturing, the tendencies of pro-frackers represent the rhetorical strategies of others who support the fossil fuel industry. Similar arguments may be found when we look to the issue of climate change. Those who resist changes to global energy portfolios often support those claims with the same kind of resilience arguments. They claim it is arrogant of humans to believe we could so significantly impact our planet's climate. To them, the earth is resilient and the greater good of supplying low-cost energy on a global scale is worth the risk. By not zooming into the local instances of climate change impact or into fracking well sites, they avoid the details that would counter their position.

While pro-frackers metaphorically provide an aerial view of the land, anti-frackers take their viewers on-site, to the ground. They tend to represent places on a smaller scale, in more detail and in sharper focus. They expound on exactly what they have witnessed, from the machinery in use to the sounds and odors of this industrial activity. By invoking formerly pristine and healthy natural places that have been subject to industry characterized as dirty and unsafe, they point to the negative transformation of place. At times they claim that these losses affect entire communities, but more than pro-frackers, they also represent these changes as negative for themselves alone. This ubiquitous NIMBY appeal especially applies to anti-frackers who lament anticipated changes to their homes. In some cases, anti-frackers even threaten to leave their communities if their side does not win.

Though such NIMBY appeals are inherently personal, anti-frackers have a strong tendency to represent places objectively while pro-frackers are noted to insert themselves into the scenes they depict. As I show in chapter 4, this is a strategy of image repair in that anti-frackers are often characterized as overly emotional and blind to scientific reasoning. By representing places void of themselves as subjects and using science-like discourse, they create scenes that are more closely aligned with scientific observation. However, their close-up view of place is amenable to NIMBY appeals in that the world is divided into many uniquely valuable places. The closer we are to a place, the more valuable it is. Pairing proximity with scientific reasoning is a characteristic rhetorical appeal of those who speak out in defense of the natural world.

A strategy common to both anti- and pro-frackers is the use of vicarious proximity, which provides insight into perceptions of risks related to economics, health and the environment. Speakers use these appeals to create a sense of proximity between audience and place, as if the audience member had experienced the place for themselves. These appeals rest on the assumption that places themselves have rhetorical power (Endres & Senda-Cook, 2011; Prelli, 2006). They assume that if an audience member could experience a place for themselves, the speaker's and audience member's perspectives on the issue would align. Drawing on the persuasive power of place, they attempt to strengthen place identity with a person not yet familiar with that place. Though I have studied vicarious proximity in the setting of a public hearing, this appeal is not limited to this context. Whether or not vicarious proximity is created differently from site to site warrants further investigation. One instance of such investigation has been carried out by Scarff (2021) on public comments gathered in response to Mountain Valley Pipeline construction in West Virginia and Virginia. In this work, Scarff finds that vicarious proximity maps onto her case study well. She also finds that holistic and dualistic metaphors are a common feature of these appeals that may even warrant the inclusion of an additional variable to vicarious proximity appeals (p. 7).

Further work in this area could also examine the use of vicarious proximity in other public settings and at other levels of government. As of yet, I am unsure to what degree vicarious proximity is used by politicians or other institutional representatives. These appeals may vary as they shift from controversy to controversy or site to site. Likewise, whether vicarious proximity is bound to land-based interventions and the public discourse that circulates in response to them requires further research.

My findings also provide explanation of identification research in mass communication that proposes, "individuals' perceptions about fracking risks appear to vary according to their psychological distance from specific threats . . . and personal health risks may be seen as more temporally and spatially proximate than environmental ones" (Feenstra et al., 2016, p. 331). Vicarious proximity provides insight into one process of creating psychological proximity to fracking risks. Beyond making an identification move, vicarious proximity appeals advance deliberative argument by strategically depicting scenes in ways that advance their argumentative purpose. While both sides of the issue use vicarious proximity at nearly equal levels, pro-frackers depict larger geographic areas in less detail while anti-frackers focus in on geographically smaller areas. These differences reinforce the widely held belief that places are always contested, being made and re-made in discourse for specific rhetorical effects (Lefebvre, 1991, p. 13). Anti-frackers wish to create a heightened sense of risk and accomplish this by close-up views and zoomed-in scenes; pro-frackers attempt to create a sense of calm, security and wide benefit by zooming out and sparing detail for coverage. How my findings extend our understanding of contests over place is the subject of the next section.

PLACE-BASED RHETORIC: THEMES, CHALLENGES, AND FINDINGS

As a contribution to rhetorical analysis, I offer renewed focus on place-based rhetoric. "Place and space" studies have benefited from wide attention, though much recent work has been on "place-as-rhetoric" (Endres & Senda-Cook, 2011, p. 258). Scholars who conceive of place-as-rhetoric find rhetorical agency in nonlinguistic artifacts and inanimate objects. I do not suggest otherwise, but my interest has instead been in how places are constructed in discourse. Though the Allegheny County Courthouse (where many of the speeches I have analyzed were given) has rhetorical power and like other places may enact rhetorical agency (Dickinson, 2002; Dickinson et al., 2010; Endres & Senda-Cook, 2011; Topinka, 2012), I show how speakers rhetorically transport their audience from their physical surroundings to places that are described in discourse. From the discourse-based angle, the power of place lies at the intersection of language, place description, and audience experience.

Even when places are not invoked with the specificity and detail to allow for audience imagining, place-in-discourse is a fruitful lens for rhetorical and discourse analysts alike. The second chapter of this book shows how focus on concrete and general places provides a part-to-whole or whole-subsumes-parts argumentative frame that encourages one view of place while obscuring others. My study of place-based discourse also offers vicarious proximity, rhetorically constructed proximity in local civic discourse, which aligns place-based discourse with narratives that emphasize setting over characters or action. The value of this category of place-based discourse is its facility to divide speakers based on their representations of place.

My contribution in this sense is methodological; I offer a heuristic analysis that starts with the places invoked within a text and broadens to the arguments constructed through place-based discourse. Representations of space, focusing on concrete and general places, allow analysts a starting point to break a text into smaller parts. Further categorization based on attitudes toward place, deliberative purpose, relative proximity to audience and place, as well as other categorizations, may provide insights, as they have in this book. In texts like those I have examined that are focused on a land-use controversy, place is a lens that provides insight into how people represent places while remaining interconnected with myriad social and argumentative purposes. However, texts concerning land use are certainly not the limit of this heuristic.

In the beginning of this book, I posed a methodological question: "What insights does a focus on place provide rhetorical analysts?" Supported by my analysis in the preceding chapters, I present perspectives on the importance of place in rhetorical analysis. I begin by focusing on the substance of place-based rhetoric. At its core, place-based rhetoric requires an explicit focus on a physical place as a component of argument. Place-based appeals are created and differentiated by discourse elements such as specificity, vividness of description, proximity, deixis, zooms, pans and scans. By analyzing the place-based appeals in public hearing speeches, it is clear that places can be represented very differently, but people whose beliefs align tend to have similar techniques. Methodologically speaking, place representation can be a tool that allows us to differentiate segments of a public and see through their eyes. In any case, this is not so simple as typical descriptions suggest: that pro-frackers see the land as a "natural resource" and anti-frackers see it as part of a larger ecology that needs to be preserved (Killingsworth & Palmer, 1992). Attention to the construction of place-based appeals from a discourse analytic perspective allows analysts to better understand how micro-rhetorical features contribute to different representations.

My analysis shows that pro- and anti-frackers invoke similar places but use different representational strategies to achieve their intended rhetorical effects. Examining how two ideologically distinct groups represent a similar place can uncover distinctions not just in their reasons for supporting a specific side, but in their assumptions about audience value hierarchies, normative ideals of humans' relationship to nature, and risk assessment. The representation of frack sites is a common tactic of both sides. By not only looking at the larger argument that this use of place supports, but interrogating these appeals on a more granular level, new insights are made available. Comparing the relative proximity in which a speaker situates their audience, for instance, allows analysts to see that it is important for some people to make their audience feel close enough to a frack site to hear, smell and see a drill operating. Others would present the same frack site from a distance, representing the whole site as a single unit. This difference matters. Like Piotr Cap's work that emphasizes the sense of impending threat created when an entity from the outside encroaches upon the deictic center (Cap, 2013), I show that close proximity is used to rhetorically create a sense of threat. What differentiates my finding from Cap's is that I show how a sense of threat is created by bringing an audience closer to the entity, not by making the entity seem as if it were coming closer to the audience.

Rather than viewing the audience as a stable (though still heterogeneous) entity, fixed to a geospatial location, my findings encourage rhetorical analysts to view the audience as mobile. Rhetors are not only presenting topics as closer or further from their audience, they are moving their audience closer to or further away from those topics. This is especially the case when the topic is place. My evidence shows that speakers go beyond considering their subject from the point of view of readers, as Karlyn Kohrs Campbell suggests (1982, pp. 149–150); they attempt to supplant their own visual perspective into the consciousness of their audience. When we consider how ideologically different speakers attempt this, their own audience analysis may account for some of the difference. For instance, when rhetors use vivid depiction to create a sense of proximity, they assume that if their audience could see what they have seen, they would be persuaded. In other words, they suggest that *seeing is believing*. The most interesting part of studying place-based appeals is determining which scenes rhetors believe will most convincingly support their arguments. The findings of my previous chapters have much to do with how near or far the audience/viewer is situated in relation to the place that is being represented. Speakers often use close proximity to create heightened threat and distance to increase a sense of safety. These strategies for presenting place at varying degrees of relative proximity to the audience not only impact an audience's perception of a place; these representational strategies also affect a speaker's ethos.

In his examination of the etymology of ethos, Arthur B. Miller (1974) defines it as a "habitual gathering place" or an "accustomed place" (p. 310). This harkens back to the early days of democracy when being a regular at the Agora demonstrated one's public character. Halloran (1982) ties this to the "image of people gathering together in a public place, sharing experiences and ideas" (p. 60). Since contemporary public hearings do not resemble the Agora, speakers today are constrained by modern, institutionalized norms of civic discourse. They are also likely not known by those who share this civic space with them.

Due to modern participation norms and the increasingly technocratic society we live in, scientific argument that draws on "established fact" is the preferred discourse of many participants. However, speakers also rely on narratives of place to vividly recreate imaginable scenes as a form of evidence. Isocrates argues that "arguments acquire more authority when they come from one's life than from mere words" (2001, p. 255). Although "mere words" overly minimizes the value of science-based arguments, many speakers do seem to agree that ethos is best built from appeals drawn from their experience rather than an ability to cite facts. In this sense, ethos is built from scene-making that is concrete and visually vivid.

This use of place for ethos construction highlights the complexity of place-based rhetoric as it accomplishes a "network of purposes" that both constructs and responds to the contextual and personal constraints of the rhetorical situation speakers enter. Flower (1994) describes this process in writing as a "web of intention . . . created by the goals, plans, criteria, desires and possibilities evoked by a specific rhetorical situation—a construction that itself develops in response to a developing text and responding readers" (p. 45). The place-based rhetoric in public hearings may at once create a scene to establish one's ethos (built on being in a place) and advance a deliberative argument about what should happen to that place, as well as other similar places. At the same time that speakers advance deliberative argument regarding the future of a place, place-based appeals can be read as epideictic in that they celebrate the importance of places in one's community. Speakers build community identity based on common concern, even in the face of disagreement. Place-based rhetoric is multifaceted and, therefore, using place as a lens for rhetorical analysis can lead to multiple insights about rhetorical intent, as we can analyze the complex constraints and responses to these rhetorical situations. In this regard, place is also considered as not just a feature of rhetorical situations, but as a rhetorical interface.

Viewing place as a rhetorical interface allows us to zoom out and consider social significance. The concept of a "rhetorical interface" comes from Jenny Rice's (2012) work on regional rhetoric. She draws from Arturo Escobar's use of the term "interface" to describe the power of ecology and economy to enable public engagement. Rice explains, "In appealing to and performing regional appeals, publics are able to address the sweeping and sometimes abstract flows of labor, food politics, migration patterns, and consumption" (p. 204). Though Rice and her co-authors focus on region (Tell, 2012; Olson, 2012; Clark, 2012; Green & Kuswa, 2012; Wood, 2012), this concept can be usefully applied to place more broadly. This is especially the case when we consider the function of place in a public hearing on impacts to the environment.

Place functions as a rhetorical interface when citizens advance competing representations that serve as evidence for arguments with social significance. Each competing representation not only shapes how audience members experience a place, such as the way anti- and pro-frackers differently create proximity to drilling operations, but these representations also serve as the nexus for a range of social issues. While issues like economic viability can be discussed without use of place as an interface, when a specific place functions as an interface, one person's concerns about ecological impact open up the door for discussion about environmental justice, economic prosperity, health, safety, and a bevy of other topics. In other words, when place is an interface, the inter-relations between social issues become more apparent as they are materially manifested in actual locations. My research shows this to be the case, but people also often reach well beyond the local.

When speakers geographically locate an issue, they "reaffirm and reclaim place" (Rice, 2012, p. 204). By representing a place in a specific way to serve a specific end, speakers attempt to control how others experience that place. This serves as an interface not only to social issues, but also to a community's value hierarchy (Perelman & Olbrechts-Tyteca, 1969). From a critical standpoint, citizens with little institutional power can affect how community places are valued, redefining them as dangerous, threatened, or unethical. DeCerteau (1984) would see this use of place as a "tactic." From this perspective, citizens who participate in public hearings are inside systems that are meant to suppress them. Still, these citizens have power. DeCerteau would say that ordinary citizens are disadvantaged by the hearing format itself, which allots them only several minutes to speak to an audience that does not verbally respond. However, DeCerteau believes that by use of tactics citizens can subvert authority. He explains that "*a way of using* imposed systems constitutes the resistance to the historical law of a state of affairs and its dogmatic relations" (p. 18). Working within the system, rhetoric is a powerful tactic, with its storehouse of tropes that can change the will of an audience (p. xx). My project in one sense has been to specify the function of place as a rhetorical tactic.

The clear lines DeCerteau sets up between the powerful and the powerless are not stable in the Marcellus Shale hearing. Place is used as a tactic by both those who support and those who oppose natural gas development. Because supporters are aligned with massively influential corporations, they may be considered powerful. However, in the Allegheny County Council chambers of March 2011, these supporters were outnumbered at least two-to-one by their (at times, very vocal) opposition. The clear power differential is between the citizen speakers and the council members at the front of the room. Speakers from both sides use place as a tactic to subvert this authority and change perceptions. Much place-based rhetoric is used to re-appropriate space in a constant contest for the right to define use. DeCerteau tells us that these speakers "*poach* . . . in countless ways on the property of others" (p. xii). Indeed, even in the case of public land use, citizens push back and forth to appropriate land for their cause. A key tactic, I have found, is the use of place-based appeals to communicate risk.

Viewed as a tactic of those with relatively little power, place-based appeals facilitate risk communication from the bottom-up. Speakers use them to provide eyewitness accounts meant to alarm community members of unseen danger, whether to ecology, human health or the economy or a foreign threat. The "contest" of the hearing is to establish which risk is the greatest. By using place as a tactic to inform and alarm community members, speakers establish their ethos as credible witnesses. Once established, they provide front-line reports to citizens and community decision-makers meant to make visible what may otherwise be overlooked.

Risk communication research is most often studied asymmetrically from the top down, even when it is socially constructed in interaction such as during debates between public figures (Buttny, 2019). Noted risk communication scholar Baruch Fischhoff explains that research has evolved to include more complicated skills and solutions to problems; however, this progress is framed as solely a venture of experts. Fischhoff (2012) explains, "Every year (or, perhaps, every day), some new industry or institution discovers that it, too, has a risk problem" (p. 208). These problems are often the type where experts identify sources of danger that are invisible to the naked eye, such as high particulate density in the air we breathe. Fischhoff provides the following developmental stages in risk management:

- All we have to do is get the numbers right.
- All we have to do is tell them the numbers.
- All we have to do is explain what we mean by the numbers.
- All we have to do is show them that they've accepted similar risks in the past.
- All we have to do is show them that it's a good deal for them.
- All we have to do is treat them nice.
- All we have to do is make them partners. (Fischhoff, 2012, p. 209)

In these stages, Fischhoff explains risk communication's development over the past 20 years, building on (but not replacing) the work of predecessors. I reproduce this development of risk communication to draw attention to the direction of communication from a rhetorical perspective. "We," in each of these stages, is always industry or institutional experts, and "them" is implied to be nonexperts who do not understand risk. Risk communication is, in other words, an enterprise from technical experts to laypeople.

Certainly, this technical expert voice is present in Marcellus Shale hearings in the form of industry experts themselves or people who cite and quote experts. However, what is most interesting about risk communication is how risk is communicated from the public to other members of the public (horizontally), or even from the public to experts of the political variety (bottom-up). Those who do not draw from scientific or technical expertise present their own experiential knowledge as expertise.

When citizens communicate risk from the bottom-up, their expertise comes from their experience with the land gained over time. They cite their familiarity with changes in the land due to fracking as a sign of heightened risk. That is, they refer to the way that fracking has transformed the landscape to show that communities are at risk. Many emphasize the long duration that they have lived in a place. Time creates a sense of expertise with the land gained by many years of observation. Although it is not necessary to communicate risk from the bottom-up, many of those who choose to couch their arguments in place-based observation do emphasize transformations they have seen over many years.

Despite the fact that speakers effectively describe changes in a place over time, many seem to miss the mark by expanding or contracting beyond the context of the hearing. Due to this, they may not tap into the real power of the county councils before them. Many speakers focus on risk that is not immediately relevant to the function of the public hearing, for instance, by emphasizing impact on an international scale. The hearings I have studied are exclusively focused on the county level. County councils are gathering public comment to make decisions about what to do in a particular county and they have power to change the course of hydraulic fracturing in their specific counties. Nevertheless, speakers often communicate risk that is either well beyond the scope of the hearing or that is personal. For instance, pro-frackers are prone to discuss the national security threat of Middle Eastern oil. Since fracking is a way to reduce our dependence on foreign resources, they claim counties should welcome the industry. Anti-frackers tend to focus on their own personal problems with the fracking industry, from expositions on personal medical problems to claims that they will move away if the risk becomes too great. In both cases, speakers often fail to situate risk at the county level. They imply that county action could aid national interests or alleviate personal problems, but often fail to focus risk communication on the county itself.

PUBLIC PARTICIPATION AND PLACE: WHAT WE LEARN

The ordinary democracy approach begins with description carefully worked through textual analysis, which enables insights that lead to normative claims. Suggestions for future practice are borne from actual practice (Craig & Tracy, 1995). Using this approach, Tracy (2010, 2016) analyses public meeting discourse found in school board meetings and hearings on same-sex marriage. This book adds public hearings on natural resource development, which draw more scientific argument and, of course, rhetorics of place.

A major normative claim that Tracy makes in this recent work is that we should accept and embrace what she calls *democracy-appealing partisanship*. If we embrace democracy-appealing partisanship, we move away from the ideals of deliberative democracy, which call for interaction among people with competing ideals so that a real change of perspective is possible (Gastil & Levine, 2005). Accepting a partisan approach, Tracy and Hughes (2014) explain, should not be cause for alarm, as healthy democracy requires that

Figure 5.1 Harrisburg, Pennsylvania. State Capital and Susquehanna River. *Source*: Photograph by Michael Yatsko.

citizens "keep competing values alive" (p. 321). Members of a public keep competing perspectives alive even if participants do not directly engage with or consider competing values. If we see the value of this type of interaction that does not require active engagement with different values as one piece of a larger democratic social system, the value of such hearings seems more tangible. In this respect, I agree wholeheartedly with Tracy, as public hearings such as the ones I have studied do not encourage meaningful engagement with perspectives that challenge a speaker. However, through articulating these perspectives, building on what others have said, and circulating the discourse that enters into and expands out from a public hearing, we may see value in this democratic practice. Community members may find new meaning in their shared places.

Nonetheless, the "democracy-appealing" aspect of this concept is not borne out in my research. Tracy provides explanation:

> To advocate while attending to the logic of the situation, citizens espouse their commitment to democracy and imply that those favoring the opposite position are being undemocratic. To state it another way, citizens use the strategy of democracy-appealing partisanship to manage the speaking challenges they face in public hearings. (Tracy & Hughes, 2014, pp. 319–320)

In a case of personal rights, as in Tracy's study on same-sex marriage hearings, claims about minority rights versus majority rule make sense. Personal rights are important in the issue of hydraulic fracturing related to the rights to develop one's private property as well as rights to clean air and water, which is assured to citizens through Pennsylvania's State Constitution. However, the public hearings I have studied are much more about economics, the environment, and public health than they are about personal rights. Democracy-appealing partisanship is not dominant. This is not to say that people do not appeal to democratic principles, but these appeals may be more useful for those studying personal rights than they are for those studying controversies over land use.

Tracy and Hughes (2014) claim that democracy-appealing partisanship "is a good normative model for doing democracy when: (1) there is a contentious issue about which many people have strong sentiments, (2) there have been few opportunities for citizens to speak to elected officials, and (3) the disputed issue is one society regards as a value-based, either-or decision rather than a multifaceted, technically complex one" (p. 321). Applied to fracking, conditions (1) and (2) are undoubtedly characteristics of the rhetorical situation participants enter. Condition (3) is less clear, but I would argue that citizens do conceive of the issue as "either-or" and value based. Of course, fracking is technically complex and its implications are multifaceted, but above all, participants boil the issue down to clear pro- and anti- sides that rest upon different valuations of economic opportunity and national security versus risk ecology, personal health and safety. If readers accept that condition (3) is present in Marcellus Shale public hearings, then I must disagree that democracy-appealing partisanship is a normative model that would be useful in all public meetings that fit these conditions. I agree that the term fits in issues about personal rights, but in hearings focused on land use, a revised normative model is better.

In disputes over land use, rather than appealing to democracy, citizens should appeal to the shared places of their communities in ways that represent salient uses and meanings of such places. Though some speakers create place-based appeals that draw from locations outside of their communities to make comparisons, the appeals that appear most effective are vivid descriptions of places that transform inert geographic locations into rhetorical interfaces that enable citizens to engage with the myriad social issues that affect and are affected by use of land. Vicarious proximity appeals are one way of accomplishing this, but other approaches that engage with concrete places are also effective, such as, but not limited to the following: scientific reasoning, historical narrative, economic analysis, and political science. Appeals that focus on concrete places rather than abstract concepts have the benefit of vividly engaging with multiple audiences in the act of co-constructing communal places.

To composition researchers and experts on professional communication, the power of vivid description should be no surprise. We know that when done well, "writing can seem sufficiently striking in visual immediacy to cause the reader to interpret the reading experience as a perceptual experience" (Kaufer & Butler, 2000, p. 86). In a real sense, words become images. The power of images aligns with a principle known to professional and technical communicators as the *picture superiority effect*. In his book *Presentation Zen,* Garr Reynolds (2008) explains that "pictures are remembered better than words, especially when people are casually exposed to the information and the exposure is for a very limited time" (p. 132). Though the rhetorical situation of a businessperson presenting a sales pitch and a citizen arguing to a county council are quite different, the picture superiority effect holds. No citizen will be able to project slides during a presentation, but they can vividly create scenes that invite the audience to imagine. Vivid depiction is the next best thing.

The Bigger Picture: The Interplay of the Global and the Local

From the national and international perspective, the issue of hydraulic fracturing in Pennsylvania and other states is much less a controversy and much more a foregone conclusion. Even though he was receptive to the local problems associated with fracking, in his 2015 State of the Union speech former US President Barack Obama univocally described the benefits of domestic energy production:

> We believed we could reduce our dependence on foreign oil and protect our planet. And today, America is number one in oil and gas. America is number one in wind power. Every three weeks, we bring online as much solar power as we did in all of 2008. (Applause.) And thanks to lower gas prices and higher fuel standards, the typical family this year should save about $750 at the pump. (Applause) (Obama, 2015)

Certainly, keeping energy affordable and reducing dependence on foreign oil and gas are both admirable goals for the United States. The lower gas prices Obama offered American citizens had much to do with domestic production of natural gas. Clearly, hydraulic fracturing has been an issue of great national and international importance. However, the local importance of fracking cannot be overshadowed and decisions about fracking should not be taken as foregone conclusions made by those seeking higher goals. As philosopher Adam Briggle (2015) explains, "Fracking represents both our global interdependence on networks of commodities and our place-bound lives in communities" (p. 48). In this sense, fracking is a good case study to understand the social interplay of the global and the local. Additionally, place is a lens that can reveal insights about this interplay due to the abstract and amorphous nature of multinational energy corporations and other players in international energy networks, contrasted with the concrete and rather stable locations and people affected by resource development.

The local perspective is particularly important in the fracking issue due to the surprising power local governments have in determining whether and how hydraulic fracturing is conducted in county or city limits. Local governments are able to stop rich and powerful energy companies in their tracks. Briggle explains that to county council members in Denton, Texas, fracking is about "zoning, flood easements, neighborhood integrity plans, plats, and permits," not about "climate change and energy independence" (p. 48). The power of county council members lies in these local decisions. Speakers at a public hearing have the opportunity to influence politicians into making decisions that can greatly impact fracking in their own communities, even if those decisions do not have much impact on the larger issues of energy independence and climate change.

However, speakers at public hearings on fracking, at least those that I have studied, very often cannot help but comment on the national and international importance of their side of the issue. I believe this is a natural response to the representative democracy in which American citizens reside, especially within our globalized world. Citizens do not have direct contact with higher levels of government beyond exceptional circumstances, yet their lives are impacted greatly by the decisions of large-scale government and business. Their attention is also influenced by media focus on national and international issues. Due to this, speakers address problems or benefits of fracking that go well beyond the power of the councils that sit before them. This desire to speak to larger social issues may make many speakers less effective than they could be. Nonetheless, focusing on issues that go beyond the scope of the hearing may also be meant to empower local government officials into believing themselves to be the small-town agents of large-scale social change. I cannot say this line of argument is wholly ineffective, but if the goal of speakers is to most persuasively influence council decisions, scope is important. Speakers do have an argumentative resource at their disposal to connect the global and the local. If you have read this far, you will easily guess that this argumentative resource is the place-based appeal.

Though we may transport audiences through rhetorical strategies like the vicarious proximity appeal, when possible we should also encourage physical experiences in places of public importance. Citizens should be able to visit hydraulic fracturing sites and other places of environmental concern. These places should be carefully documented, observed, and experienced. When industry restricts access, as is often the case, other means may be possible.

Phaedra Pezzullo (2009) describes such an experience in Louisiana's "Cancer Alley" among other sites. This kind of "toxic tourism" allows more privileged citizens to see, hear, smell, and feel what fellow community members survive every day. This can also raise awareness more broadly to the lived experiences of people in environmental justice communities. By being bodily present in these places, we can form regional bonds across communities of concern through more deeply informed place-based appeals.

Ecology, Place, and Students

At the end of this book, we understand the importance of place-based rhetoric for citizens engaging in land-based disputes. Many of the speakers whose words we have studied gained their aptitude for place through long hours spent in the places they described. Their vicarious proximity appeals draw from their personal experience. They do at times miss the mark in terms of scope, reaching well beyond the geographic influence of the councils before them or so narrowly defining impact that their words come across as personal grievance more than argument for the common good. So, then, how can we teach our students in rhetoric, composition, and communication classes how to effectively argue from place?

Many people, not just young people, lack contact with the natural world. Some see the need for vicarious representations of place as problematic in and of itself. Responding to societal shifts toward technocracy, Killingsworth (2007) explains, "At the heart of the problem lies the human capacity to forget where we actually are, to live virtually or vicariously, to lose one's connections to other people and the natural habitat" (p. 59). This worry of people becoming detached from place, especially the natural world, is recognized in popular literature (Louv, 2005), and reflected in the concept of "biophilia," which portrays humans as innately connected to nature through evolutionary history (Wilson, 1984). Vicarious proximity appeals do not directly address this lack of the natural world, but they do speak to the potential energy of these experiences. When we lead students to meaningful experiences in the outdoors, they can begin to make connections between place, society, and the future they wish to inhabit. They must develop connections with the outside world, but they also need exposure to the language that effectively communicates affective ties to land. As we have seen, we communicate volumes when, as objective observers or subjective guides, we locate the objects our discourse in varying degrees of proximity to us in the deictic center. We express group affiliations, push against stereotypes, advance our own hierarchies of value, and commit them to memory.

Certainly, students must be able to convincingly describe the places that they encounter so that audiences can visualize their words. Nature writers and environmental advocates like John Muir, Edward Abbey, Terry Tempest Williams, Annie Dillard, and many others can drive home this point. A line like "The sky is deep and distant, laced with sycamore limbs like a hatching of crossed swords" (Dillard, 1974/2013, p. 101) may not find its way into a public hearing, but the visual imagery that overflows from a book like *Pilgrim at Tinker Creek* can inspire students to not only see the world more clearly, but to creatively present it. But how did Dillard come to see the sky and the sycamores in this way? Just like the participants in public hearings that make place-based appeals, she spent a long time observing this place. For her, Tinker Creek is more than a place; it's an interface, a nexus. She comes to see Tinker Creek ecologically in the broadest sense of the word, connecting its physical and material elements to the social, historical, theological, and philosophical. To teach students how to effectively communicate about places, we need to teach them skills of close observation paired with the ability to look beyond our immediate surroundings to connect to larger spheres of concern in the world. Ideally, we provide them with experiences of "land affect," or nonmediated exposure to the natural world that help them "feel with the land" (Arnold, 2018, p. 97). These experiences need not be dramatic, as I have learned from teaching students to write from place, despite Arnold's argument that such connections are "seldom registered in modern human awareness until the power of the land is mobilized and brought to bear upon embodied subjectivities" (p. 97). To Arnold, major disruptions like natural disasters or other deeply affective experiences express the power of the land. What I see from students is that such severity is not necessary to form this bond. Precisely because we have such a thirst for the natural world due to biophilia, many people *can* form these connections to a place through repeated exposure, research, and focus.

One way to do this is through an assignment called the *Tiny Ecology Project*, which asks students to immerse themselves in a place and all its myriad connections. I have used this project to help students develop place-based experience and skills in arguing on behalf of places. In Jeffrey Cohen's (2013) original version of this project, he helps students see the "surfacings of particular histories" that convene on any place and develop in students' "ecological attentiveness." To do this, students select a small and (preferably) unassuming location to study for the duration of a semester. They choose short stretches of streamside banks, planter boxes outside of their dorms, a particular bench in a park, or a corner of their backyard. They journal about this place, noting what catches their eye with particular attention paid to change and interaction. Not only is this a site to closely observe; they must also research what they find there. Cohen (2013) explains, "you may have to learn the difference between kudzu and dandelion; a starling and a wren." They may also need to investigate the history of that place. How did it come to be? What was it before? Students study these places and post about them in bi-weekly discussion board posts.

My revision of the Tiny Ecology Project differs from Cohen's in that I ask students to go beyond closely observing and researching what they find. This added step is that they must use the place as the ground from which they produce rhetoric. Students must also present the communal value of their place to the class. Most everyone has little trouble becoming personally connected to these places simply through repeated visits and the attention paid to them, but the challenge they face—the same challenge faced by fracking public hearing participants—is making what is of personal concern an issue of broader import. *We can hear that you care, but why should this place matter to the rest of us?* Answering that question requires rhetorical acuity and strategic research. In a sense, this is similar to Ackerman's (2007) *georhetorical method*, which helps students not only see, but "struggle against the historical and material precedents" that connect the writer to their community (p. 114). These students must reach beyond the personal to address the values tied to these places and the people who inhabit them. Doing this requires students to look at their sites as not just one geospatially fixed location, but as a representative place.

Taking the perspective that place is representative helps students connect the value they find in their tiny ecologies with the likely interests and values of their audience. In the culminating assignment of this project, they vividly depict places, bearing witness to what they have found and evaluating the interactions that converge on the site. They connect this to their audience by identifying similar sites in the community and reaching beyond the place they have occupied. Still, this also can be issue focused. For instance, students often find invasive species in their sites, which lead them to research interspecies conflicts that are as much ideological as they are biological. Should we value more highly what is native? Does Japanese knotweed (non-native to our area) deserve to die? What about the large and goo-filled bug, the spotted lanternfly (also non-native to our area)? By grounding their perspective on ideological disputes like these in their tiny ecologies, students write with the first-hand knowledge that helps them convince their audiences to pay attention. My hope is that students recall this experience if they should find themselves facing a county council, arguing on behalf of the land, or in any of the other ways they may act as local citizens.

This assignment is adaptable to many contexts, learning outcomes, student levels, and periods of time. I run this project in both my first-year writing courses as well as an upper-level topics course titled Environmental Advocacy Writing. In the lower-level course, I emphasize the similarities between the typical research and writing process and the study of a tiny ecology. Students find useful and memorable the connections between place-as-research site and later research projects more typical in nature. In a sense, we draw on the power of spatial memory to support their learning of the research and writing process. In the upper-level course, our focus shifts to the depiction of place, the intertwining threads of biology, sociology, history, geology, and politics, as well as threats to the environment. Both versions of the project lead the best students to truly insightful, nuanced arguments that visually represent overlooked places as valuable to audiences. Students use vicarious proximity appeals to show us a part of the world that means a lot to them; they help us see why it matters. Even students who are not inclined to give their full effort report developing stronger observational skills and connections to their particular place. I take heart from that in these days of digital consumption.

Beyond this, such projects can also work as partnerships between campus and the local community. Students may observe public spaces, engage with their community, and document this experience to be shared publicly. From his perspective of critical regionalism that celebrates local difference, Powell explains:

> Whatever forces act to isolate academia from the local scene, however, colleges and universities over one very powerful opportunity for social change: access to students, and to the opportunity to teach them ways to invent their own descriptions of the world. So critical regionalism must be, ultimately, a pedagogy, one that teaches students how to draw their own regional maps connecting their experience to that of others near and far, both like and unlike themselves. (Powell, 2007, p. 7)

The Tiny Ecology Project is one such experience that can help students bridge this gap between their worlds and the wider world around them, but other scholars are also developing critical regionalist pedagogies (Brown, 2019).

Opportunities for students to learn outside of our classroom walls have always been important. Ancient Roman rhetorician, Cicero, remarkably said that "he owed less to the schools of the rhetoricians than to the gardens of the Academy. Nor indeed would so wonderful a fertility of mind have displayed itself in him, if he had circumscribed his genius by the limits of the forum, and not allowed it to range through all the domains of nature" (Bizzell & Herzberg, 2001, p. 421). This is not to say we should send our students away to learn from nature alone, though quite likely they would gain from the experience. What I am saying here is that if we want our students to engage with places as rhetorical interfaces and to be able to transfer that skill when they need it later in life, we need to provide opportunities for them to practice in the unrulier domains of nature. This is especially important as contact with greenspaces has been shown to improve spatial memory (Flouri et al., 2019).

Drawing on our human penchant for place and memory, we can activate in learners connections to place that can serve as the locus of rhetoric. We may foster this simply by holding class outdoors on a campus green, through the Tiny Ecology Project, or by more involved field experiences that are typically considered the domain of the sciences.

Rhetorical fieldwork is an inclusive term for the kind of engagement humanities and social science scholars attuned to strategic communication can use to collect more robust and pervasive data about the sites they study (Rai, 2016). In their book titled *Field Rhetoric*, Rai and Gottschalk-Druschke (2018) explain that this movement toward fieldwork responds to a "shift toward studying local, ordinary, vernacular, deeply situational, and spatial manifestations of rhetoric (and writing) that emerge within everyday life" (p. 7). Such an approach helps us study place from the inside, viewing it from the ground where we can best find a strand to follow the inevitable networks that exist there. This perspective and the one I encourage my students to take as they study their tiny ecologies is more holistic, more ecological than the approach taken in this book. The Weyekin Principle, field rhetoric, rhetorical ecologies, and other ecological approaches to rhetorical situations are deepened by close textual analysis. This book remains focused on a type of text, the public hearing speech, and analyzes the public rhetoric using a range of textual methods from sentence-level discourse analysis to the corpus-level view afforded by computer-aided analysis. But this book remains focused on texts more traditionally defined and does not venture to account for an ecological view of fracking in its various locations. As a result, this book does not directly leverage the insights offered by new materialism, but I argue that discourse analyses such as this one can substantiate and extend the possibilities of ecological studies of rhetoric by providing empirical grounding for their claims. For an example of this, I turn to another book on fracking in Pennsylvania: Jimmy Guignard's *Pedaling the Sacrifice Zone* (2015). Guignard takes an embodied approach to rhetorical scholarship as he examines his own life in the "sacrifice zone" of Tioga County, Pennsylvania. His cycling trips provide Guignard with an embodied presence in spaces that became industrial zones as the natural gas industry moved in. A trained rhetorician, Guignard connects his lived experience and careful attention to the natural world with the changing social reality and symbolic representations of it. Guignard explains, "Whether on the truck or on my bike, I experience a small slice of physical reality that is mediated by the language, the symbols, I've encountered at home, at school, at work, one the news, wherever via whatever media" (p. 57). This experientially focused approach is just the kind of ecological perspective espoused by Salvador and Clarke in their article on the Weyekin Principle, discussed in the Introduction.

What especially strikes me about Guignard's work is the overlapping findings between our two studies. For instance, Guignard's ecological approach has him reviewing a broader range of texts including media and industry representations. About this, Guignard writes:

> Taken collectively, the stories demonstrate how the industry is talked about in an abstract sense when things are going well, like when land is leased or mergers are made, yet when pollution occurs, it happens to specific people in particular places. The rhetoric used by the industry mystifies its relationship with the place it drills. It's not that jobs or lease money are bad, but they are transcendent and apply to any place. (Guignard, p. 82)

His finding that the industry (pro-frackers) relies on abstract language that is less tied to places corresponds to my own findings about the representations of place that are characteristic of industry supporters. Likewise, Guignard's insights about pollution happening to "specific people in particular places" relate to the representations of place by anti-frackers that bring audiences close. However, Guignard also reveals this is a tactic used by industry supporters to isolate pollution events. This is one example of many where Guignard's findings and my own overlap, which helps substantiate the claims that we both make.

Where Guignard's work provides a broader, contextually rich take on the rhetoric of hydraulic fracturing, my work drills deep through layers of language in use. That we reach the same conclusions should build confidence in both approaches. As rhetoricians take more ecological approaches to sites of study, we should continue to enlist the insights of textually focused analysts who make use of increasingly powerful approaches to deciphering the intricacies of language in use. Likewise, textual analysts should look to ecological approaches to see how they fit into the network of connections between their texts and the worlds they inhabit.

This book has shown that citizens—through their use of place-based appeals as they serve as a rhetorical interface—have the power to ground expansive social issues in the real-world impact on concrete communities, people, and places. How citizens formulate these appeals to place differs greatly as they articulate their ideological position, relationship to the issue, and argumentative purpose. Citizens are afforded multiple paths between themselves, their communities, and the ways their communities connect with the larger world, while those of us who study rhetoric and discourse are afforded a lens through which to understand these connections.

Bibliography

Aakhus, M., & Lewiński, M. (2017). Advancing polylogical analysis of large-scale argumentation: Disagreement management in the fracking controversy. *Argumentation*, *31*(1), 179–207.

Ackerman, J. A. (2007). Teaching the capital city. In C. J. Keller & C. R. Weisser (Eds.). *The Locations of Composition.* Albany: State University of New York Press.

Ackerman, J. A., Druschke, C. G., McGreavy, B., & Sprain, L. (2016). the Skunkwork of Ecological Engagement. *Reflections*, *16*(1), 75–95.

Applegarth, R. (2011). Genre, location, and Mary Austin's ethos. *Rhetoric Society Quarterly*, *41*(1), 41–63.

Arnold, J. (2018). Feeling the fires of climate change: Land affect in Canada's Tar Sands. In K. Bladow (Ed.). *Affective Ecocriticism.* Lincoln: University of Nebraska Press.

Asen, R. (2003). Imagining in the public sphere. *Philosophy and Rhetoric*, *35*(4), 345–367.

Asen, R. (2004). A discourse theory of citizenship. *Quarterly Journal of Speech*, *90*(2), 189–211.

Auburn, T., & Barnes, R. (2006). Producing place: A neo-Schutzian perspective on the 'psychology of place'. *Journal of Environmental Psychology*, *26*(1), 38–50.

Avraham, E. (2016). Destination marketing and image repair during tourism crises: The case of Egypt. *Journal of Hospitality and Tourism Management*, *28*, 41–48.

Bamberg, M., & Georgakopoulou, A. (2008). Small stories as a new perspective in narrative and identity analysis. *Text & Talk: An Interdisciplinary Journal of Language, Discourse Communication Studies*, *28*(3), 377–396.

Barber, B. (1984). *Strong democracy: Participatory politics for a new age.* Berkeley and Los Angeles: University of California Press.

Barton, E. (2002). Resources for discourse analysis in composition studies. *Style*, *36*(4), 575–594.

Barton, E., & Stygall, G. (Eds.). (2002). *Discourse studies in composition.* Cresskill, NJ: Hampton Press.

Baudrillard, J. (1975). *The mirror of production* (Vol. 17). St. Louis: Telos Press.

Beck, U. (2003). *Risk society: Towards a new modernity*. London: Sage.

Bednarek, M., & Caple, H. (2010). Playing with environmental stories in the news—good or bad practice?. *Discourse & Communication*, *4*(1), 5–31.

Begos, K. (2013, October 22). Marcellus Shale gas growing faster than expected. *Houston Chronicle: Fuel Fix*. Retrieved from: http://fuelfix.com/blog/2013/10/22/marcellus-shale-gas-production-continues-to-grow/

Benhabib, S. (1992). Models of public space: Hannah Arendt, the liberal tradition, and Jürgen Habermas. In C. Calhoun (Ed.). *Habermas and the public sphere* (73–98). Cambridge, MA: MIT Press.

Benhabib, S. (1996). Toward a deliberative model of democratic legitimacy. In S. Benhabib (Ed.). *Democracy and difference: Contesting the boundaries of the political* (67–94). Princeton, NJ: Princeton University Press.

Benoit, W. L. (1995). Sears' repair of its auto service image: Image restoration discourse in the corporate sector. *Communication Studies*, *46*(1–2), 89–105.

Benoit, W. L. (1997). Image repair discourse and crisis communication. *Public Relations Review*, *23*(2), 177–186.

Benoit, W. L. (2000). Another visit to the theory of image restoration strategies. *Communication Quarterly*, *48*(1), 40–43.

Benoit, W. (2017). Image repair on the Donald Trump "Access Hollywood" video: "Grab them by the p* ssy". *Communication Studies*, *68*(3), 243–259.

Benoit, W. L., & Drew, S. (1997). Appropriateness and effectiveness of image repair strategies. *Communication Reports*, *10*(2), 153–163.

Benoit, W., Holtzhausen, D., & Zerfass, A. (2015). Image Repair Theory in the context of the Strategic Communication. In D. R. Holtzhausen & A. Zerfass (Eds.). *Routledge Handbook of Strategic Communication* (303–311). New York: Routledge.

Biber, D. (1986). Spoken and written textual dimensions in English: Resolving the contradictory findings. *Language*, *62*(2), 384–414.

Bickerstaff, K., & Walker, G. (2003). The place (s) of matter: matter out of place–public understandings of air pollution. *Progress in Human Geography*, *27*(1), 45–67.

Bierwisch, M. (1996). How much space gets into language? In P. Bloom, M. Peterson, L. Nadel, & M. Garrett (Eds.). *Language and space* (31–76). Cambridge, MA: MIT Press.

Bitzer, L. (1968). The rhetorical situation. *Philosophy and Rhetoric*, *1*(1), 1–14.

Bizzell, P., & Herzberg, B. (2001). *The rhetorical tradition: Readings from classical times to the present*. Boston: Bedford/St. Martin's.

Blommaert, J., Dewilde, A., Stuyck, K., Peleman, K., & Meert, H. (2003). Space, experience and authority: Exploring attitudes towards refugee centers in Belgium. *Journal of Language and Politics*, *2*(2), 311–331.

Bormann, E. G. (1972). Fantasy and rhetorical vision: The rhetorical criticism of social reality. *Quarterly journal of Speech*, *58*(4), 396–407.

Brasch, W. M. (2013). *Fracking Pennsylvania: Flirting with disaster*. Carmichael, CA: Greeley & Stone.

Brasier, K. J., Filteau, M. R., McLaughlin, D., Jacquet, J., Stedman, R. C., Kelsey, T. W., & Goetz, S. J. (2011). Residents' perceptions of community and environmental impacts from development of natural gas in the Marcellus Shale: A comparison of Pennsylvania and New York cases. *Journal of Rural Social Sciences*, *26*, 32–61.

Briggle, A. (2015). *A field philosopher's guide to fracking: How one Texas town stood up to big oil and gas.* New York: Liverlight Publishing Company.

Brown, L. M. (2019). Flyover states, No Man's Land, and the Bible Belt: Introducing critical regionalism for rhetorical analysis. *Communication Teacher*, *34*(3), 1–6.

Brown, P., & Levinson, S. C. (1987). *Politeness: Some universals in language use.* Cambridge: Cambridge University Press.

Bucholtz, M., & Hall, K. (2005). Identity and interaction: A sociocultural linguistic approach. *Discourse Studies*, *7*(4–5), 585–614.

Bühler, K. (1990). *Theory of language: The representational function of language* (Vol. 25). Amsterdam: John Benjamins Publishing.

Burke, K. (1969). *A grammar of motives.* Berkeley: Univ of California Press.

Bush, J., Moffatt, S., & Dunn, C. (2001). 'Even the birds round here cough':: stigma, air pollution and health in Teesside. *Health & Place*, *7*(1), 47–56.

Buttny, R. (2015). Contesting hydrofracking during an inter-governmental hearing: Accounting by reworking or challenging the question. *Discourse & Communication*, *9*(4), 423–440.

Buttny, R. (2019). Debating hydrofracking: The discursive construction of risk. *Frontiers in Communication.* 4(5), 1–10. *doi: 10.3389/fcomm.*

Buttny, R., & Cohen, J. R. (2007). Drawing on the words of others at public hearings: Zoning, Wal-Mart, and the threat to the aquifer. *Language in Society*, *36*(5), 735–756.

Buttny, R & Feldpausch-Parker, A. M. (2016). Communicating Hydrofracking. *Environmental Communication, 10*(3), 289–291.

Buttny, R., & Williams, P. L. (2000). Demanding respect: The uses of reported speech in discursive constructions of interracial contact. *Discourse & Society*, *11*(1), 109–133.

Calhoun, C. (1993). Civil Society and the Public Sphere. *Public Culture*, *5*(2), 267–280.

Campbell, K. K. (1982). *The rhetorical act.* Belmont: Wadsworth Pub Co.

Cantrill, J. G., & Masluk, M. D. (1996). Place and privilege as predictors of how the environment is described in discourse. *Communication Reports*, *9*(1), 79–84.

Cantrill, J. G., & Oravec, C. L. (Eds.). (1996). *The symbolic earth: Discourse and our creation of the environment.* Lexington: University Press of Kentucky.

Cap, P. (2013). *Proximization: The pragmatics of symbolic distance crossing.* Amsterdam: John Benjamins Pub. Company.

Cap, P. (2018). 'We don't want any terrorists here': The linguistic manufacturing of xenophobia in the post-2015 Poland. *Discourse & Society*, *29*(4), 380–398.

Carbaugh, D. (1996). *Situating selves: The communication of social identities in American scenes.* Albany: SUNY Press.

Carbaugh, D. (1999). "Just listen": "Listening" and landscape among the Blackfeet. *Western Journal of Communication (includes Communication Reports)*, *63*(3), 250–270.

Carbaugh, D. (2007). Quoting "the environment": Touchstones on earth. *Environmental Communication, 1*(1), 64–73.

Carbaugh, D., & Cerulli, T. (2013). Cultural discourses of dwelling: Investigating environmental communication as a place-based practice. *Environmental Communication, 7*(1), 4–23.

Chen, S., & Gunster, S. (2016). 'Ethereal carbon': Legitimizing liquefied natural gas in British Columbia. *Environmental Communication, 10*(3), 305–321.

Cheng, M. S. (2012). Colin Powell's speech to the UN: A discourse analytic study of reconstituted ethos. *Rhetoric Society Quarterly, 42*(5), 424–449.

Chilton, P. (2004). *Analyzing Political Discourse: Theory and Practice.* London: Routledge.

Chilton, P. (2005). Discourse space theory. *Annual Review of Cognitive Linguistics, 3*, 78–116.

Clark, G. (2004). *Rhetorical landscapes in America: Variations on a theme from Kenneth Burke*. Univ of South Carolina Press.

Clark, G. (2012). "A child born of the land": The rhetorical aesthetic of Hawaiian song. *Rhetoric Society Quarterly, 42*(3), 251–270.

Cloud, D. L. (1994). The materiality of discourse as oxymoron: A challenge to critical rhetoric. *Western Journal of Communication (includes Communication Reports), 58*(3), 141–163.

Cloud, D. (2019). The rise of the gay warrior: Rhetorical archetypes and the transformation of identity categories. *Discourse & Communication, 13*(1), 26–47.

Cohen, J. J. (2013). Ecologies of conquest/contact ecologies [Blog post]. *In the Medieval Middle.* Retrieved from http://www.inthemedievalmiddle.com/2013/08/ecologies-of-conquest-contact-ecologies.html)

Committee on Industry, Research and Energy. (2015, September 18). Opinion of the committee on the Committee on the Environment, Public Health and Food Safety. *Report on Towards a European Energy Union.* European Parliament. A8-0341/2015.

Coombs, W. T., Frandsen, F., Holladay, S. J., & Johansen, W. (2010). Why a concern for apologia and crisis communication? *Corporate Communications: An International Journal, 15*(4), 337–349.

Craig, R. T., & Tracy, K. (1995). Grounded practical theory: The case of intellectual discussion. *Communication Theory, 5*(3), 248–272.

Cramer, P. A. (2011). *Controversy as news discourse* (Vol. 19). Dordrecht: Springer Science & Business Media.

Cronon, W. (1995). *Uncommon ground: Toward reinventing nature*. New York: Norton.

Cresswell, T. (2013). *Place: A short introduction*. Malden, MA: Blackwell Publishing.

Crowley, S., & Hawhee, D. (1999). Ancient rhetorics for contemporary students. Boston: Allyn and Bacon.

De Certeau, M., & Rendall, S. (1984). *The practice of everyday life*. Berkeley: University of California Press.

Delaware River Basin Commission. (2021, February 25). New DRBC regulation prohibits High Volume Hydraulic Fracturing in the Delaware River Basin [Press release]. Retrieved from: https://www.state.nj.us/drbc/home/newsroom/news/approved/20210225_newsrel_HVHF-rulemaking.html

Depoe, S. P.. (1998). Talking about the Earth: On the Growing Significance of Environmental Communication Studies [Review of *Ecocritique: Contesting the Politics of Nature, Economy, and Culture; Sharing the Earth: The Rhetoric of Sustainable Development; Thinking Ecologically: The Next Generation of Environmental Policy*]. *Rhetoric and Public Affairs, 1*(3), 435–448. Retrieved from http://www.jstor.org/stable/41939463

Dewey, J. (1927). *The public and its problems*. New York: H. Holt and Company.

Dickinson, G. (2002). Joe's rhetoric: Finding authenticity at Starbucks. *Rhetoric Society Quarterly, 32*(4), 5–27.

Dickinson, G., Blair, C., & Ott, B. L. (2010). *Places of public memory: The rhetoric of museums and memorials*. Tuscaloosa: University of Alabama Press.

Dillard, A. (2013). *Pilgrim at Tinker Creek*. New York: Harper Perennial Modern Classics. (Original in 1974).

Dobrin, S. I., & Morey, S. (2009). *Ecosee: Image, rhetoric, nature*. Albany: SUNY Press.

Dokshin, F. A. (2016). Whose backyard and what's at issue? Spatial and ideological dynamics of local opposition to fracking in New York State, 2010 to 2013. *American Sociological Review, 81*(5), 921–948.

Downey, S. D. (1993). The evolution of the rhetorical genre of apologia. *Western Journal of Communication (includes Communication Reports), 57*(1), 42–64.

Druschke, C. G. (2013). Watershed as common-place: Communicating for conservation at the watershed scale. *Environmental Communication, 7*(1), 80–96.

Duggan-Hass, D., Ross, R. M., & Allom, W. D. (2013). The science beneath the surface: A very short guide to the Marcellus Shale. Ithaca, NY: Paleontological Research Institution.

Dunmire, P. (2011). *Projecting the future through political discourse: The case of the Bush doctrine*. Amsterdam, Netherlands: John Benjamins.

Edbauer, J. (2005). Unframing models of public distribution: From rhetorical situation to rhetorical ecologies. *Rhetoric Society Quarterly, 35*(4), 5–24.

Einstein, K. L., Palmer, M., & Glick, D. M. (2019). Who participates in local government? Evidence from meeting minutes. *Perspectives on Politics, 17*(1), 28–46.

Emerson, R. W. (1986). *Emerson on transcendentalism*. Bloomsbury Publishing.

Endres, D., & Senda-Cook, S. (2011). Location matters: The rhetoric of place in protest. *Quarterly Journal of Speech, 97*(3), 257–282.

Entrikin, J. N. (1991). *The betweenness of place*. Houndmills: Macmillan.

Environmental Protection Agency. 4Fraking42412-920. Retrieved from https://www.epa.gov/sites/production/files/hfstudy/images/4Fraking42412-920.jpg

Fahnestock, J. (2011). *Rhetorical style: The uses of language in persuasion*. Oxford: Oxford University Press.

Fairclough, N. (1989). *Language and power*. London: Longman.

Farbotko, C. (2005). Tuvalu and climate change: Constructions of environmental displacement in the Sydney Morning Herald. *Geografiska Annaler: Series B, Human Geography*, *87*(4), 279–293.

Farkas, K. R. (2013). Power and access in the public hearings of city council meetings. *Discourse & Society*, *24*(4), 399–420.

Farrell, T. B., & Goodnight, G. T. (1981). Accidental rhetoric: the root metaphors of Three Mile Island. *Communication Monographs*, *48*(4), 271–300.

Feenstra, A. S., Lyons, B. A., & Fowler-Dawson, A. (2016). Conservatism vs. conservationism: Differential influences of social identities on beliefs about fracking. *Environmental Communication*, *10*(3), 322–336.

Fischhoff, B. (2012). *Risk analysis and human behavior*. Abingdon, Oxon: Earthscan.

Flouri, E., Papachristou, E., & Midouhas, E. (2019). The role of neighbourhood greenspace in children's spatial working memory. *British Journal of Educational Psychology*, *89*(2), 359–373.

Flower, L. (1994). *The construction of negotiated meaning: A social cognitive theory of writing*. Carbondale: Southern Illinois University Press.

Flower, L., & Hayes, J. R. (1984). Images, plans and prose: The representation of meaning in writing. *Written Communication*, *1*(1), 120–160.

Foucault, M., Gordon, C., Marshall, L., Mepham, J., & Soper, K. (1980). *Power / knowledge: Selected interviews and other writings, 1972–1977*. New York: Vintage.

Fox, J., Gandour, M., Adlesic, T., Sanchez, M., International WOW Company, Gasland Productions., HBO Documentary Films, New Video Group. (2010). *Gasland: Can you light your water on fire?* New York, NY: Docurama Films.

Frampton, K. (1998). Towards a critical regionalism: Six points for an architecture of resistance. In H. Foster (Ed.). *The anti-aesthetic: Essays on postmodern culture* (17–34). New York: The New Press.

Fraser, N. (1990). Rethinking the public sphere: A contribution to the critique of actually existing democracy. In C. Calhoun (Ed.). *Habermas and the public sphere* (109–142). Cambridge, MA: MIT Press.

Gastil, J., & Levine, P. (2005). *The deliberative democracy handbook: Strategies for effective civic engagement in the twenty-first century*. San Francisco: Jossey-Bass.

Geisler, C. (2018). Coding for language complexity: The interplay among methodological commitments, tools, and workflow in writing research. *Written Communication*, *35*(2), 215–249.

Glaser, B. G., & Strauss, A. L. (2009). *The discovery of grounded theory: Strategies for qualitative research*. New Brunswick, NJ: Aldine Transaction.

Gold, R. (2014). *The boom: How fracking ignited the American energy revolution and changed the world*. New York: Simon & Schuster.

Golden, J. L., Corbett, E. P. J., Blair, H., Campbell, G., & Whately, R. (1990). *The Rhetoric of Blair, Campbell, and Whately*. Carbondale: Southern Illinois University Press.

Greene, R. W., & Kuswa, K. D. (2012). "From the Arab Spring to Athens, From Occupy Wall Street to Moscow": Regional Accents and the Rhetorical Cartography of Power. *Rhetoric Society Quarterly*, *42*(3), 271–288.

Guignard, J. S. (2015). *Pedaling the sacrifice zone: Teaching, writing, and living above the Marcellus shale*. College Station: Texas A&M University Press.

Habermas, J. (1989). *The structural transformation of the public sphere: An inquiry into a category of bourgeois society*. Cambridge, Mass: MIT Press.

Habermas, J., & Cronin, C. (2015). *The lure of technocracy*. Cambridge: Polity.

Halliday, M. A. K., & Martin, J. R. (2003). *Writing science: Literacy and discursive power*. London: Taylor & Francis.

Halloran, S. M. (1982). Aristotle's Concept of Ethos, or If Not His Somebody Else's. *Rhetoric Review, 1*(1), 58–63.

Harlow, W. F., Brantley, B. C., & Harlow, R. M. (2011). BP initial image repair strategies after the Deepwater Horizon spill. *Public Relations Review*, *37*(1), 80–83.

Hart, R. P. (2015). Genre and Automated Text Analysis: A Demonstration. In J. Ridolfo & W. H. Davidson (Eds.). *Rhetoric and the Digital Humanities* (152). Chicago: University of Chicago Press.

Haspel, K., & Tracy, K. (2007). Marking and shifting lines in the sand: Discursive moves of ordinary democracy. In Tracy, K., McDaniel, J. P., & Gronbeck, B. E. (Eds). *The prettier doll: Rhetoric, discourse, and ordinary democracy* (142–175). University of Alabama Press.

Hauser, G. A. (1999). *Vernacular voices: The rhetoric of publics and public spheres*. Columbia: University of South Carolina Press.

Hearit, K. M. (1995). "Mistakes were made": Organizations, apologia, and crises of social legitimacy. *Communication Studies*, *46*(1–2), 1–17.

Hearit, K. M. (2018). Corporate Apologia Organizational Rhetoric and Wrongdoing. In O. Ihlen & R. Heath (Eds.). *The Handbook of Organizational Rhetoric and Communication* (185). Hoboken: Wiley Blackwell.

Heffernan, J. A. W. (1993). Museum of words: The poetics of ekphrasis from Homer to Ashbery. Chicago: University of Chicago Press.

Hesford, W. S. (2004). Documenting violations: Rhetorical witnessing and the spectacle of distant suffering. *Biography*, *27*(1), 104–144.

Howarth, R., Ingraffea, A., & Engelder, T. (2011). Should fracking stop? *Nature*, 477, 271–275.

Hopey, D. (2010, July 23). 1,200 hear Marcellus Shale debate. *Pittsburgh Post-Gazette*. Retrieved from http://www.post-gazette.com/local/washington/2010/07/23/1-200-hear-Marcellus-Shale-debate/stories/201007230146

Huckin, T. N. (1992). Context-sensitive text analysis. In G. Kersch & P. Sullivan (Eds.). *Methods and Methodology in Composition Research* (84–104). Carbondale: Southern Illinois University Press.

Huckin, T. (1997). Critical discourse analysis. In T. Miller (Ed.). *Functional approaches to written text* (78–92). Washington, DC: U.S. Information Agency.

Huckin, T., Andrus, J., & Clary-Lemon, J. (2012). Critical discourse analysis and rhetoric and composition. *College Composition and Communication*, *64*(1), 107–129.

Hunsaker, D. M., & Smith, C. R. (1973). Rhetorical distance: A critical dimension. *Western Speech*, *37*(4), 241–252.

Ishizaki, S., & Kaufer, D. (2011). Computer-aided rhetorical analysis. In P. M. McCarthy & C. Boonthum-Denecke (Eds.). *Applied natural language processing: Identification, investigation, and resolution* (276–296). Hershey, PA: IGI Global.

Isocrates. (2001). Antidosis. In Bizzell, P & Herzberg, B. (Eds.). *The Rhetorical Tradition: Readings from Classical Times to the Present* (75–80). Bedford: St. Martin's.

Jansa, K. (Producer). (2011, March 10). Allegheny County Public Hearing on Marcellus Shale. Video retrieved from https://vimeo.com/21280329.

Jarratt, S. (1998). Ekphrastic Rhetoric and National Identity in Adam Smith's Rhetoric Lectures. In L. L. Gaillet (Ed.). *Scottish Rhetoric and Its Influences* (43–56). Mahwah, N.J: Hermagoras Press.

Jeffery, K. J. (2018). The hippocampus: from memory, to map, to memory map. *Trends in Neurosciences*, *41*(2), 64–66.

Johnstone, B. (2011). *Discourse analysis*. Malden, MA: Blackwell Publishing.

Johnstone, B., Andrus, J., & Danielson, A. E. (2006). Mobility, indexicality, and the enregisterment of "Pittsburghese". *Journal of English linguistics*, *34*(2), 77–104.

Johnstone, B., & Eisenhart, C. (2008). *Rhetoric in detail: Discourse analyses of rhetorical talk and text*. Amsterdam: John Benjamins.

Johnstone, B., & Mando, J. (2015). Proximity and journalistic practice in environmental discourse: Experiencing 'job blackmail' in the news. *Discourse & Communication*, *9*(1), 81–101.

Jones, R. H. (2011). Data collection and transcription in discourse analysis. In B. Paltridge & K. Hyland (Eds.). *Continuum companion to discourse analysis* (9–21). London: Continuum.

Kalin, J., & Frith, J. (2016). Wearing the city: Memory p (a) laces, smartphones, and the rhetorical invention of embodied space. *Rhetoric Society Quarterly*, *46*(3), 222–235.

Katz, S. B., & Miller, C. R. (1996). The low-level radioactive waste siting controversy in North Carolina: Toward a rhetorical model of risk communication. In C. G. Herndl & S. Brown (Eds.). *Green culture: Environmental rhetoric in contemporary America* (111–140). Madison: University of Wisconsin Press.

Kaufer, D. S., & Butler, B. S. (2000). *Designing interactive worlds with words: Principles of writing as representational composition*. New York: Routledge.

Kaufer, D., Geisler, C., Ishizaki, S., & Vlachos, P. (2005). Textual genre analysis and identification. In Y. Cai (Ed.). *Ambient intelligence for scientific discovery* (129–151). Berlin: Springer.

Kaufer, D., & Hariman, R. (2008). Discriminating political styles as genres: A corpus study exploring Hariman's theory of political style. *Text & Talk-An Interdisciplinary Journal of Language, Discourse Communication Studies*, *28*(4), 475–500.

Killingsworth, M. J. (2007). A Phenomenological Perspective on Ethical Duty in Environmental Communication. *Environmental Communication*, *1*(1), 58–63.

Killingsworth, M. J., & Palmer, J. S. (1992). *Ecospeak: Rhetoric and Environmental Politics in America*. Carbondale: Southern Illinois University Press.

Kjeldsen, J. E. (2008). Mediated publics and rhetorical fragmentation. In *Democracy, Journalism and Technology: New Developments in an Enlarged Europe* (115–129). Tartu: Tartu University Press.

Koban, J. (2019). Ecological restoration or healing?: Conflicting ontologies and missed opportunities in public debates surrounding Mississippi River Gorge restoration. *Environmental Communication*, *14*(5), 1–15.

Kroepsch, A. (2016). New rig on the block: spatial policy discourse and the new suburban geography of energy production on Colorado's Front Range. *Environmental Communication, 10*(3), 337–351.

Labov, W. (1972). *Language in the inner city: Studies in the Black English vernacular.* Philadelphia: University of Pennsylvania Press.

Labov, W., Waletzky, J., & Helm, J. (1967). Narrative analysis: Oral versions of personal experience. In J. Helm (Ed.). *Essays on the verbal and visual arts.* Seattle: University of Washington Press.

Laclau, E., & Mouffe, C. (1985). *Hegemony and socialist strategy: Towards a radical democratic politics.* London: Verso.

Lakoff, G. (2010). Why it matters how we frame the environment. *Environmental Communication, 4*(1), 70–81.

Larson, G. S., & Pearson, A. R. (2012). Placing identity: Place as a discursive resource for occupational identity work among high-tech entrepreneurs. *Management Communication Quarterly, 26*(2), 241–266.

Lefebvre, H. (1991). *The production of space* (Vol. 142). Blackwell: Oxford.

Levasseur, D. G., & Carlin, D. B. (2001). Egocentric argument and the public sphere: citizen deliberations on public policy and policymakers. *Rhetoric & Public Affairs, 4*(3), 407–431.

Lynch, D. A. (1998). Rhetorics of proximity: Empathy in Temple Grandin and Cornel West. *Rhetoric Society Quarterly, 28*(1), 5–23.

McGee, M. C. (1990). Text, context, and the fragmentation of contemporary culture. *Western Journal of Communication, 54*(3), 274–289.

McGuire, M. (1990). The rhetoric of narrative: A hermeneutic, critical theory. In B. Britton & A. D. Pelligrini (Eds.). *Narrative thought and narrative Language* (219–236). New York: Psychology Press.

McKerrow, R. E. (1989). Critical rhetoric: Theory and praxis. *Communications Monographs, 56*(2), 91–111.

McKerrow, R. E. (1991). Critical rhetoric in a postmodern world. *Quarterly Journal of Speech, 77*(1991), 75–78.

Mando, J. (2016). Constructing the vicarious experience of proximity in a Marcellus Shale public hearing. *Environmental Communication, 10*(3), 352–364.

Mando, J., & Stack, G. (2019). Convincing the public to kill: Asian carp and the proximization of invasive species threat. *Environmental Communication, 13*(6), 820–833.

Mansbridge, J. J. (1980). *Beyond adversary democracy.* New York: Basic Books.

Marsh, C. (2006). The syllogism of apologia: Rhetorical stasis theory and crisis communication. *Public Relations Review, 32*(1), 41–46.

Matthews, J., & Hansen, A. (2018). Fracturing debate? A review of research on media coverage of "fracking". *Frontiers in Communication, 3*(41).

Matz, J., & Renfrew, D. (2015). Selling "fracking": Energy in Depth and the Marcellus Shale. *Environmental Communication, 9*(3), 288–306.

Metze, T., & Dodge, J. (2016). Dynamic discourse coalitions on hydro-fracking in Europe and the United States. *Environmental Communication, 10*(3), 365–379.

Miller, A. B. (1974). Aristotle on habit (εθō) and character (ηθō): Implications for the rhetoric. *Speech Monographs, 41*(4), 309–316.

Milstein, T., Anguiano, C., Sandoval, J., Chen, Y. W., & Dickinson, E. (2011). Communicating a "new" environmental vernacular: A sense of relations-in-place. *Communication Monographs*, *78*(4), 486–510.

Modan, G. (2002). 'Public toilets for a diverse neighborhood': Spatial purification practices in community development discourse. *Journal of Sociolinguistics*, *6*(4), 487–513.

Moore, M. P. (1993). Constructing irreconcilable conflict: The function of synecdoche in the spotted owl controversy. *Communications Monographs*, *60*(3), 258–274.

Muir, J. (1908). The Hetch Hetchy Valley. *Sierra Club Bulletin*, *6*(4), 212–220. Retrieved from: http://vault.sierraclub.org/ca/hetchhetchy/hetch_hetchy_muir_scb_1908.html

Myers, G. (2006). 'Where are you from?': Identifying place. *Journal of Sociolinguistics*, *10*(3), 320–343.

Newman, J. H., & DeLaura, D. J. (1968). *Apologia pro vita sua*. New York: W.W. Norton & Company, Inc.

Obama, B. H. (2015, January). *State of the Union Address*. Speech presented at Washington, D.C. Retrieved from: https://www.whitehouse.gov/the-press-office/2015/01/20/remarks-president-state-union-address-january-20-2015.

Ochs, E., & Capps, L. (1997). Narrative authenticity. *Journal of Narrative and Life History*, *7*, 83–90.

Oddo, J. (2013). Discourse-based methods across texts and semiotic modes: Three tools for micro-rhetorical analysis. *Written Communication*, *30*(3), 236–275.

Oddo, J. (2014). *Intertextuality and the 24-hour news cycle: A day in the rhetorical life of Colin Powell's U.N. address*. East Lansing: Michigan State University Press.

Ogneva-Himmelberger, Y., & Huang, L. (2015). Spatial distribution of unconventional gas wells and human populations in the Marcellus Shale in the United States: vulnerability analysis. *Applied Geography*, *60*, 165–174.

Olson, C. J. (2012). "Raíces Americanas": Indigenist art, América, and arguments for Ecuadorian nationalism. *Rhetoric Society Quarterly*, *42*(3), 233–250.

Ono, K. A., & Sloop, J. M. (1992). Commitment to telos—A sustained critical rhetoric. *Communications Monographs*, *59*(1), 48–60.

Oravec, C. (1981). John Muir, Yosemite, and the sublime response: A study in the rhetoric of preservationism. *Quarterly Journal of Speech*, *67*(3), 245–258.

Osborn, M. (2004). Rhetorical distance in letter from Birmingham Jail. *Rhetoric & Public Affairs*, *7*(1), 23–35.

Peeples, J. (2011). Toxic sublime: Imaging contaminated landscapes. *Environmental Communication*, *5*(4), 373–392.

Perelman, C., & Olbrechts-Tyteca, L. (1969). The new rhetoric: A treatise on argumentation. Notre Dame, IN: University of Notre Dame Press.

Perry, S. (2013). "Strange fruit," Ekphrasis, and the lynching scene. *Rhetoric Society Quarterly*, *43*(5), 449–474.

Pezzullo, P. C. (2009). *Toxic tourism: Rhetorics of pollution, travel, and environmental justice*. Birmingham: University of Alabama Press.

Phillips, K. R. (2010). The failure of memory: Reflections on rhetoric and public remembrance. *Western Journal of Communication*, *74*(2), 208–223.

Phillips, S. (2011, September 7) "Fractivists" vs. "Factivists": Chesapeake CEO says opponents have facts wrong. *State Impact Pennsylvania: National Public Radio.* Retrieved from: https://stateimpact.npr.org/pennsylvania/2011/09/07/fractivists-vs-factivists/

Physicians for Social Responsibility & Concerned Health Professionals of New York. (2019, July). *Compendium of scientific, medical, and media findings demonstrating risks and harms of fracking.* Retrieved from: https://www.psr.org/blog/resource/compendium-of-scientific-medical-and-media-findings-demonstrating-risks-and-harms-of-fracking/

Powell, D. R. (2007). *Critical regionalism: Connecting politics and culture in the American landscape.* Chapel Hill: University of North Carolina Press Books.

Prasch, A. M. (2015). Reagan at Pointe du Hoc: Deictic epideictic and the persuasive power of "bringing before the eyes". *Rhetoric & Public Affairs, 18*(2), 247–275.

Prasch, A. M. (2015). Retelling Watergate: Apologia, Political Eulogy, and Richard Nixon's "Final Campaign". *Southern Communication Journal, 80*(4), 271–292.

Prelli, L. J. (2006). *Rhetorics of display.* Columbia, SC: University of South Carolina Press.

Rai, C. (2016). *Democracy's lot: Rhetoric, publics, and the places of invention.* Tuscaloosa: The University of Alabama Press.

Rai, C., & Druschke, C. G. (2018). *Field rhetoric: Ethnography, ecology, and engagement in the places of persuasion.* Tuscaloosa: The University of Alabama Press.

Raimi, D. (2017). *The fracking debate: The risks, benefits, and uncertainties of the shale revolution.* New York: Columbia University Press.

Rasmussen, S. G., Ogburn, E. L., & McCormack, M., et al. (2016). Association Between Unconventional Natural Gas Development in the Marcellus Shale and Asthma Exacerbations. *JAMA Internal Medicine, 176*(9), 1334–1343.

Reynolds, G. (2008). *Presentation zen: Simple ideas on presentation design and delivery.* Berkeley, CA: New Riders Pub.

Rhetorica Ad Herennium. (1954). *Trans. Harry Caplan.* Cambridge: Harvard UP.

Rich, J. L. (2016). Drilling is just the beginning: Romanticizing rust belt identities in the campaign for shale gas. *Environmental Communication, 10*(3), 292–304.

Rice, J. (2012). From architectonic to tectonics: Introducing regional rhetorics. *Rhetoric Society Quarterly, 42*(3), 201–213.

Riley, C. L. (2015). When cows talk: The happy California cow campaign as visual apologia. *Argumentation and Advocacy, 51*(4), 273–289.

Roberts-Miller, P. (2004). *Deliberate conflict: Argument, political theory, and composition classes.* Carbondale: Southern Illinois University Press.

Rogers, R. A. (1998). Overcoming the objectification of nature in constitutive theories: Toward a transhuman, materialist theory of communication. *Western Journal of Communication, 62*(3), 244–272.

Salvador, M., & Clarke, T. (2011). The Weyekin principle: Toward an embodied critical rhetoric. *Environmental Communication: A Journal of Nature and Culture, 5*(3), 243–260.

Scarff, K. (2021). Mapping vicarious proximity: Holistic and dualistic metaphors in a Mountain Valley Pipeline public hearing. *Environmental Communication*, OnlineFirst, 1–16.

Schegloff, E. (1972). Notes on a conversational practice: Formulating place. In D. Sudnow (Ed.). *Studies in social interaction*. New York: Free Press. 259–279.

Schiappa, E. (2003). *Defining reality: Definitions and the politics of meaning*. Carbondale: Southern Illinois University Press.

Schiffrin, D. (2009). Crossing boundaries: The nexus of time, space, person, and place in narrative. *Language in Society*, *38*(4), 421–445.

Seegert, N. (2014). Queer beasts: Ursine punctures in domesticity. *Environmental Communication*, *8*(1), 75–91.

Senda-Cook, S. (2013). Materializing tensions: How maps and trails mediate nature. *Environmental Communication*, *7*(3), 355–371.

Sernovitz, G. (2016). *The green and the black: The complete story of the shale revolution, the fight over fracking, and the future of energy*. New York: St. Martin's Press.

Simons, H. (2007). Political performances in public proceedings: the social dramas of Barbiegate. In K. Tracy, J. P. McDaniel, & B. E. Gronbeck (Eds). *The prettier doll: Rhetoric, discourse, and ordinary democracy* (45–69). Tuscaloosa: University of Alabama Press.

Smith, M. (1995). *Literary realism and the ekphrastic tradition*. University Park, PA: Pennsylvania State University Press.

Smith, M. B., Montgomery, C. T. (2015). *Hydraulic fracturing*. Boca Raton: CRC Press/Taylor & Francis Group.

Subramanian, M. (2019, Winter). United in change. *Orion Magazine,* 38(4). Retrieved from: https://orionmagazine.org/article/united-in-change/.

Supreme Court of Pennsylvania. (2020, February 27). The Forty-third statewide investigating grand jury. *71 W.D. MISC. DKT. 2017. Allegheny County Common Pleas CP-02-MD-5947-2017. Notice No. 42.*

Swales, J. (1990). *Genre analysis: English in academic and research settings*. Cambridge: Cambridge University Press.

Takach, G. (2013). Selling nature in a resource-based economy: Romantic/extractive gazes and Alberta's bituminous sands. *Environmental Communication: A Journal of Nature and Culture*, *7*(2), 211–230.

Tannen, D. (2007). *Talking voices: Repetition, dialogue and imagery in conversational discourse*. Cambridge: Cambridge University Press.

Tell, D. (2012). The meanings of Kansas: Rhetoric, regions, and counter regions. *Rhetoric Society Quarterly*, *42*(3), 214–232.

Thompson, M. (2013). *Institutional Rhetoric, Argument, and the House Un-American Activities Committee, 1955–1956*. (Doctoral dissertation). Carnegie Mellon University.

Topinka, R. J. (2012). Resisting the fixity of suburban space: The walker as rhetorician. *Rhetoric Society Quarterly*, *42*(1), 65–84.

Tracy, K. (2010). *Challenges of ordinary democracy: A case study in deliberation and dissent*. University Park: Pennsylvania State University Press.

Tracy, K. (2016). *Discourse, Identity, and Social Change in the Marriage Equality Debates*. Oxford: Oxford University Press.
Tracy, K., & Hughes, J. M. (2014). Democracy-appealing partisanship: A situated ideal of citizenship. *Journal of Applied Communication Research*, *42*(3), 307–324.
Tracy, K., McDaniel, J. P., & Gronbeck, B. E. (2007). *The prettier doll: Rhetoric, discourse, and ordinary democracy*. Tuscaloosa: University of Alabama Press.
Tuan, Y. F. (1977). *Space and place: The perspective of experience*. Minneapolis: University of Minnesota Press.
Tversky, B. (2003). Structures of mental spaces how people think about space. *Environment and Behavior*, *35*(1), 66–80.
Tzonis, A., & Lefaivre, L. (1981). The grid and the pathway. *Architecture in Greece*, *15*, 164–178.
Usher, M. (2013). Defending and transcending local identity through environmental discourse. *Environmental Politics*, *22*(5), 811–831.
Van Dijk, T. A. (1985). *Discourse and communication: New approaches to the analysis of mass media discourse and communication.* Berlin: W. de Gruyter.
Van Dijk, T. A. (1990). Discourse & Society: a new journal for a new research focus. *Discourse & Society*, *1*(1), 5–16.
Van Leeuwen, T. (2008). *Discourse and practice: New tools for critical discourse analysis*. Oxford: Oxford University Press.
Vivian, B. (2018). Memory: Ars Memoriae, Collective Memory, and the Fortunes of Rhetoric. *Rhetoric Society Quarterly*, *48*(3), 287–296.
Volkmer, I. (2008). Conflict-related media events and cultures of proximity. *Media, War & Conflict*, *1*(1), 90–98.
Ware, B. L., & Linkugel, W. A. (1973). They spoke in defense of themselves: On the generic criticism of apologia. *Quarterly Journal of Speech*, *59*(3), 273–283.
Warner, M. (2010). *Publics and counterpublics*. New York: Zone Books.
Watts, J. (2011, April 21). China takes step towards tapping shale gas potential with first well. *The Guardian.* Retrieved from: https://www.theguardian.com/environment/2011/apr/21/china-shale-gas-well
Wilber, T. (2012). *Under the surface: Fracking, fortunes and the fate of the Marcellus Shale*. Ithaca, NY: Cornell University Press.
Williams, L., Macnaghten, P., Davies, R., & Curtis, S. (2017). Framing 'fracking': Exploring public perceptions of hydraulic fracturing in the United Kingdom. *Public Understanding of Science*, *26*(1), 89–104.
Wilson, J. M., O'Leary, M., Metiu, A., & Jett, Q. R. (2008). Perceived proximity in virtual work: Explaining the paradox of far-but-close. *Organization Studies*, *29*(7), 979–1002.
Wodak, R. (1989). *Language, power, and ideology: Studies in political discourse*. Amsterdam: J. Benjamins Pub. Co.
Wood, A. (2012). Regionalization and the construction of ephemeral co-location. *Rhetoric Society Quarterly*, *42*(3), 289–296.
Young, I. M. (1996). Communication and the other: Beyond deliberative democracy. In S. Benhabib (Ed.). *Democracy and difference: Contesting the boundaries of the political* (120–135). Princeton, NJ: Princeton University Press.

Zelizer, B. (1990). Where is the author in American TV news? On the construction and presentation of proximity, authorship, and journalistic authority. *Semiotica*, *80*(1–2), 37–48.

Zilcosky, K. (2018, October 4). *MarcellusMoney.org report reveals millions in donations to state elected officials from the fracking industry* [Press Release] Retrieved from: https://www.conservationpa.org/news/marcellusmoneyorg-report-reveals-millions-donations-state-elected-officials-fracking-industry

Zhang, Y., Gautam, R., Pandey, S., Omara, M., Maasakkers, J. D., Sadavarte, P., . . . & Zhang, R. (2020). Quantifying methane emissions from the largest oil-producing basin in the United States from space. *Science Advances*, *6*(17).

Index

About the Author

Justin Mando is an associate professor of science and technical writing in the Department of English at Millersville University of Pennsylvania. He completed his PhD in rhetoric at Carnegie Mellon University. His research investigates the intersection of rhetoric of place, environmental controversy, and public discourse. His research has been published in *Environmental Communication*, *Discourse & Communication*, and *Composition Studies*. He lives with his wife and two children in York, Pennsylvania, where he spends his free time exploring the forests and streams. He volunteers as the chair of the Board of the Lower Susquehanna Riverkeeper Association.

www.ingramcontent.com/pod-product-compliance
Ingram Content Group UK Ltd.
Pitfield, Milton Keynes, MK11 3LW, UK
UKHW020841290926
15470UKWH00004B/1488

9 781793 62087